RESIDUE REVIEWS

VOLUME 83

RESIDUE REVIEWS

RESIDUE REVIEWS

Residues of Pesticides and Other
Contaminants in the Total Environment

Editor
FRANCIS A. GUNTHER

Assistant Editor
JANE DAVIES GUNTHER

Riverside, California

VOLUME 83

Properties, Effects, Residues, and Analytics of the
Insecticide Endosulfan

By

H. Goebel, S. Gorbach, W. Knauf, R.H. Rimpau,
and H. Hüttenbach

SPRINGER-VERLAG
NEW YORK   HEIDELBERG   BERLIN
1982

Library of Congress Catalog Card Number 62-18595.

The use of general descriptive names, trade names, trademarks, etc. in this publication, even if the former are not especially identified, is not to be taken as a sign that such names, as understood by the Trade Marks and Merchandise Marks Act, may accordingly be used freely by anyone.

New York: 175 Fifth Avenue, New York, N.Y. 10010
Heidelberg: 6900 Heidelberg 1, Postfach 105 280, West Germany

---

ISBN 978-1-4612-5714-1   ISBN 978-1-4612-5712-7 (eBook)
DOI 10.1007/978-1-4612-5712-7

Foreword

Worldwide concern in scientific, industrial, and governmental communities over traces of toxic chemicals in foodstuffs and in both abiotic and biotic environments has justified the present triumvirate of specialized publications in this field: comprehensive reviews, rapidly published progress reports, and archival documentations. These three publications are integrated and scheduled to provide in international communication the coherency essential for nonduplicative and current progress in a field as dynamic and complex as environmental contamination and toxicology. Until now there has been no journal or other publication series reserved exclusively for the diversified literature on "toxic" chemicals in our foods, our feeds, our geographical surroundings, our domestic animals, our wildlife, and ourselves. Around the world immense efforts and many talents have been mobilized to technical and other evaluations of natures, locales, magnitudes, fates, and toxicology of the persisting residues of these chemicals loosed upon the world. Among the sequelae of this broad new emphasis has been an inescapable need for an articulated set of authoritative publications where one could expect to find the latest important world literature produced by this emerging area of science together with documentation of pertinent ancillary legislation.

The research director and the legislative or administrative advisor do not have the time even to scan the large number of technical publications that might contain articles important to current responsibility; these individuals need the background provided by detailed reviews plus an assured awareness of newly developing information, all with minimum time for literature searching. Similarly, the scientist assigned or attracted to a new problem has the requirements of gleaning all literature pertinent to his task, publishing quickly new developments or important new experimental details to inform others of findings that might alter their own efforts, and eventually publishing all his supporting data and conclusions for archival purposes.

The end result of this concern over these chores and responsibilities and with uniform, encompassing, and timely publication outlets in the field of environmental contamination and toxicology is the Springer-Verlag (Heidelberg and New York) triumvirate:

> *Residue Reviews* (vol. 1 in 1962) for basically detailed review articles concerned with any aspects of residues of pesticides and other chemical contaminants in the total environment, including toxicological considerations and consequences.

Bulletin of Environmental Contamination and Toxicology (vol. 1 in
1966) for rapid publication of short reports of significant advances
and discoveries in the fields of air, soil, water, and food contamination and pollution as well as methodology and other disciplines
concerned with the introduction, presence, and effects of toxicants
in the total environment.

Archives of Environmental Contamination and Toxicology (vol. 1 in
1973) for important complete articles emphasizing and describing
original experimental or theoretical research work pertaining to the
scientific aspects of chemical contaminants in the environment.

Manuscripts for *Residue Reviews* and the *Archives* are in identical
formats and are subject to review, by workers in the field, for adequacy
and value; manuscripts for the *Bulletin* are not reviewed and are published
by photo-offset to provide the latest results without delay. The individual
editors of these three publications comprise the Joint Coordinating Board
of Editors with referral within the Board of manuscripts submitted to one
publication but deemed by major emphasis or length more suitable for
one of the others.

Coordinating Board of Editors

Preface

That residues of pesticide and other contaminants in the total environment are of concern to everyone everywhere is attested by the reception accorded previous volumes of "Residue Reviews" and by the gratifying enthusiasm, sincerity, and efforts shown by all the individuals from whom manuscripts have been solicited. Despite much propaganda to the contrary, there can never be any serious question that pest-control chemicals and food-additive chemicals are essential to adequate food production, manufacture, marketing, and storage, yet without continuing surveillance and intelligent control some of those that persist in our foodstuffs could at times conceivably endanger the public health. Ensuring safety-in-use of these many chemicals is a dynamic challenge, for established ones are continually being displaced by newly developed ones more acceptable to food technologists, pharmacologists, toxicologists, and changing pest-control requirements in progressive food-producing economies.

These matters are of genuine concern to increasing numbers of governmental agencies and legislative bodies around the world, for some of these chemicals have resulted in a few mishaps from improper use. Adequate safety-in-use evaluations of any of these chemicals persisting into our foodstuffs are not simple matters, and they incorporate the considered judgments of many individuals highly trained in a variety of complex biological, chemical, food technological, medical, pharmacological, and toxicological disciplines.

It is hoped that "Residue Reviews" will continue to serve as an integrating factor both in focusing attention upon those many residue matters requiring further attention and in collating for variously trained readers present knowledge in specific important areas of residue and related endeavors involved with other chemical contaminants in the total environment. The contents of this and previous volumes of "Residue Reviews" illustrate these objectives. Since manuscripts are published in the order in which they are received in final form, it may seem that some important aspects of residue analytical chemistry, biochemistry, human and animal medicine, legislation, pharmacology, physiology, regulation, and toxicology are being neglected; to the contrary, these apparent omissions are recognized, and some pertinent manuscripts are in preparation. However, the field is so large and the interests in it are so varied that the editors and the Advisory Board earnestly solicit suggestions of topics and authors to help make this international book-series even more useful and informative.

"Residue Reviews" attempts to provide concise, critical reviews of timely advances, philosophy, and significant areas of accomplished or needed endeavor in the total field of residues of these and other foreign chemicals in any segment of the environment. These reviews are either general or specific, but properly they may lie in the domains of analytical chemistry and its methodology, biochemistry, human and animal medicine, legislation, pharmacology, physiology, regulation, and toxicology; certain affairs in the realm of food technology concerned specifically with pesticide and other food-additive problems are also appropriate subject matter. The justification for the preparation of any review for this book-series is that it deals with some aspect of the many real problems arising from the presence of any "foreign" chemicals in our surroundings. Thus, manuscripts may encompass those matters, in any country, which are involved in allowing pesticide and other plant-protecting chemicals to be used safely in producing, storing, and shipping crops. Added plant or animal pest-control chemicals or their metabolites that may persist into meat and other edible animal products (milk and milk products, eggs, etc.) are also residues and are within this scope. The so-called food additives (substances deliberately added to foods for flavor, odor, appearance, etc., as well as those inadvertently added during manufacture, packaging, distribution, storage, etc.) are also considered suitable review material. In addition, contaminant chemicals added in any manner to air, water, soil or plant or animal life are within this purview and these objectives.

Manuscripts are normally contributed by invitation but suggested topics are welcome. Preliminary communication with the editors is necessary before volunteered reviews are submitted in manuscript form.

Department of Entomology   F.A.G.
University of California   J.D.G.
Riverside, California
March 15, 1982

RESIDUE REVIEWS

VOLUME 83

Properties, effects, residues, and analytics of the insecticide Endosulfan*

By

H. Goebel,** S. Gorbach,** W. Knauf,** R.H. Rimpau,**
and H. Hüttenbach**

Contents

General introduction 5
  I. Chemical and physical properties of endosulfan and its
    degradation products.
    By H. Goebel 6
    a) Introduction. 6
    b) Synthesis and physicochemical properties 6
        1. Chemical designation 6
        2. Synthesis of technical endosulfan 6
        3. Synthesis of radioactively labeled endosulfan 7
        4. Physicochemical properties 8
        5. Formulations 8
    c) Transformation products 9
    d) Chemical and photochemical degradation of endosulfan
       and its metabolites. 10
        1. Chemical degradation. 10
        2. Photochemical degradation 11
Summary. 12
  II. Analytical methods for endosulfan residues and metabolites.
    By S. Gorbach 13
    a) Qualitative determination 13
    b) Quantitative determination 15
        1. In formulated products and in the technical grade
          substance. 15
        2. Residue analysis 15
          α. Extraction 15
          β. Cleanup. 17
          γ. Gas-chromatographic determination 18

---

*Translated by Dr. Hugo Behr.
**Hoechst Aktiengesellschaft, Frankfurt (M)-Höchst, W. Germany.

δ. Thin-layer chromatographic separation   20
ε. High-performance liquid chromatography as
   determination method for endosulfan   21
ζ. Recommended methods for the determination of
   endosulfan residues   21
c) Analysis of endosulfan metabolites   23
   1. Introduction   23
   2. Extraction   23
   3. Cleanup   24
   4. Gas-chromatographic determination   24
   5. Thin-layer chromatographic separation
      and identification   25
   6. High-performance liquid chromatographic
      separation   26
Summary   26
III. Metabolism of endosulfan in plants and animals.
   By S. Gorbach   27
   a) Plants   27
   b) Microorganisms   28
   c) Insects   30
   d) Warm-blooded animals   30
Summary   34
IV. Toxicity of endosulfan and its metabolites.
   By R. H. Rimpau   37
   a) Introduction   37
   b) Two-year feeding test   37
   c) Three-generation reproduction study   37
   d) Teratogenicity test   37
   e) Cancerogenicity tests   38
   f) Mutagenicity test   38
   g) Acute oral toxicity of the isomers and metabolites   38
Summary   38
V. Environmental toxicology of endosulfan and its metabolites.
   By W. Knauf   38
   a) Introduction   38
   b) Aquatic organisms   39
      1. Heterotrophic microorganisms (bacteria and
         soil fungi)   39
      2. Autotrophic microorganisms   39
         α. Unicellular green algae   39
         β. Filamentous green and bluegreen algae   39
      3. Aquatic animal organisms   42
         α. Insects, crayfish, molluscs (fish-feed animals)   42
         β. Fish   44
   c) Terrestrial organisms   46
      1. Plants   47
      2. Animals   47

   α. Invertebrates 47
   β. Vertebrates 51
Summary. 52
 VI. The environmental behavior of endosulfan and residue
  values.
  By S. Gorbach 52
  a) Factors influencing the residue behavior of endosulfan 52
   1. Volatility. 52
   2. Adsorption. 54
   3. Hydrolysis 54
   4. Oxidation 55
   5. Photodegradation 56
   6. Biochemical degradation. 56
  b) Endosulfan residues in the soil 56
  c) Endosulfan (I) residues in water 57
  d) Uptake of endosulfan residues by animals and plants 71
   1. Water as environment. 71
   2. Terrestrial environment 74
   3. Laboratory ecosystem 74
  e) Residues in food and stimulants. 74
   1. Vegetable foodstuffs and forage crops. 74
    α. Apple 75
    β. Grape 75
    γ. Cabbage. 75
    δ. Tomato 75
    ε. Spinach 75
    ζ. Lettuce 76
    η. Cotton 76
    ϑ. Sugarcane 76
    ι. Sweet potatoes 76
    κ. Black currants. 76
   2. Endosulfan residues in tea and tea infusions 76
   3. Endosulfan residues in green and roasted coffee, further
    in coffee infusions. 76
   4. Endosulfan residues in tobacco 79
   5. Endosulfan residues in the daily food 83
  f) The influence on endosulfan residues by the
  processing of food and stimulants. 86
   1. Vegetable products 86
   2. Foodstuffs of animal origin 88
Summary. 88
 VII. Effect of endosulfan and its metabolites on arthropods.
  By W. Knauf. 89
  a) Introduction. 89
  b) Effect of endosulfan (I). 89
   1. Contact effect. 89
   2. Feed effect. 96

      3. Combined feed and contact effect 99
   c) Comparative effect of the single isomers of
      endosulfan (I) and its metabolites 101
   d) External intoxication symptoms. 104
   e) Physiological responses of treated insects 105
      1. Oxygen consumption 105
      2. Inner body temperature 107
      3. Effect on water turnover. 108
      4. Pulse rate. 109
      5. pH value of hemolymph 110
      6. Further reactions. 110
Summary 111
VIII. Tolerances for endosulfan.
   By R. H. Rimpau. 112
  IX. Thiodan ®: Fields of application and survey of technical
   literature.
   By. H. Hüttenbach. 112
   a) Introduction. 112
   b) Formulations 113
   c) Application rates, special notes, and tolerances 113
   d) Fields of application 120
      1. Annual field crops: Grain crops 120
         α. Wheat 120
         β. Maize (corn) and sorghum. 120
         γ. Rice 121
      2. Annual field crops: Leguminous crops. 122
         α. Beans and peas 122
         β. Soybeans. 122
      3. Annual field crops: Tuber and root crops. 123
         α. Potatoes 123
         β. Sugarbeets 123
      4. Annual field crops: Fiber and oil crops 123
         α. Cotton 123
         β. Jute 125
         γ. Rape, mustard seed, and safflower 125
         δ. Castor-oil plants 126
         ε. Sunflower 126
         ζ. Peanuts 126
      5. Annual field crops: Vegetable crops 126
         α. Tomatoes and Spanish peppers. 126
         β. Cabbages and onions 127
         γ. Okra (Abelmoschus) 127
         δ. Mushrooms 127
         ε. Carrots and parsley 128
      6. Annual field crops: Other crops. 128
         α. Tobacco 128

7. Fruit crops: Soft fruit 128
   α. Currants and blackberries 128
   β. Strawberries................................ 129
   γ. Pineapples 129
   δ. Vines 129
8. Fruit crops: Top fruit.......................... 129
   α. Apples and pears............................ 129
   β. Peaches................................... 130
   γ. Hazelnuts and walnuts 130
   δ. Mangoes 131
   ε. Date palms................................ 131
   ζ. Citrus 131
9. Plantation crops: Coffee....................... 132
10. Plantation crops: Cocoa 133
11. Plantation crops: Tea.......................... 133
12. Plantation crops: Oil palms 134
13. Plantation crops: Rubber trees................... 134
14. Plantation crops: Olives 134
15. Plantation crops: Mulberry trees................. 134
16. Plantation crops: Sugarcane..................... 134
17. Plantation crops: Pepper........................ 135
18. Forage crops: Lucerne and clover................. 135
   e) Forestry 135
   f) Other uses of Thiodan: Control of tsetse fly 137
List of pests mentioned in this publication..................... 137
References.. 141

General introduction

Endosulfan is an insecticide possessing a relatively broad spectrum of activity; as a cyclic sulfite ester, it differs from the chlorinated hydrocarbons of the cyclodiene group by its physiological properties and behavior in the organism as well as by its extremely sparing action on beneficial insects. The technical active substance consists of two isomers, α- and β-endosulfan, which both possess similar insecticidal properties. Endosulfan was synthesized at *Hoechst Aktiengesellschaft* by Frensch and Goebel (1954 and 1954 a), the insecticidal effect of which was discovered by Frensch *et al.* (1954 b), and brought onto the market under the registered trade name Thiodan®.

Maier-Bode (1968) reported in *Residue Reviews* on the literature recorded until 1967. Since that time, a host of further results, especially in the field of environment and residues, have become known. This publication gives a survey on the literature published since 1968.

I. Chemical and physical properties of endosulfan and its degradation products

By H. Goebel

a) Introduction

In order to present both a well-rounded picture of endosulfan and its metabolites as well of the chemistry of endosulfan necessary for the comprehension of the biological chapters, the chemical data largely compiled by Maier-Bode (1968) are presented hereinafter in concise form.

b) Synthesis and physicochemical properties of endosulfan

1. **Chemical designation.**—In the literature the following designations are used for endosulfan (I):

$$C_9H_6Cl_6O_3S \qquad\qquad I$$

5-norbornene-2,3-dimethanol-1,4,5,6,7,7-hexachloro-cyclic sulfite (Chemical Abstracts)

α, β-1,2,3,4,7,7-hexachlorobicyclo-[2,2,1]-heptene-(2)-bis-hydroxymethylene-(5,6)-sulfite

6,7,8,9,10,10-hexachloro-1,5,5a,6,9,9a-hexahydro-6,9,methano-2,4,3-benzo-dioxathiepin-3-oxide (IUPAC)

2. **Synthesis of technical endosulfan.**—Endosulfan (I) is synthesized by diene synthesis of hexachlorocyclopentadiene and *cis*-butene-(2)-diol-(1,4), or by hydrolysis of endosulfandiol diacetate [XIII, Diels-Alder addition product from hexachlorocyclopentadiene and *cis*-1,4-diacetoxybutene-(2)] to form endosulfandiol (II), which is caused to react with thionyl chloride. There forms a constant mixture of isomers containing approximately 70% α-endosulfan and 30% β-endosulfan (αI, βI). Either isomer yields on hydrolysis the same sterically uniform endosulfandiol (II):

Steric formulae:

II   αI   βI

The preparative separation of the endosulfan isomers (αI, βI) is feasible by column chromatography on aluminum oxide with tetrachloromethane and benzene as mobile phase (Weinmann 1970).

An analysis of the structure and confirmation of endosulfan (αI) in solid phase and in solution was described by Byrn and Siew (1977).

3. **Synthesis of radioactively labeled endosulfan.**—Various authors described the preparation of radioactively labeled endosulfan:

IA   IB   IC

IIA   IIB   IID

Korte and Stiasni (1962) report on the synthesis of II A; I A is obtained therefrom by reaction with thionyl chloride. A modified synthesis of I A and II A was carried out by Herok (1964). I B and II B form according to Frensch and Goebel (1954 and 1954 a) by using ^{14}C-hexachlorocyclopentadiene; I C may be obtained after Forman et al. (1960) by starting from ^{14}C-acetylene. ^{35}S-endosulfan (I D) is obtained by reaction of II with [35]SOCl$_2$. A further synthesis starting from ^{14}C-maleic acid anhydride was described by Huhtanen and Dorough (1978).

4. Physicochemical properties.—*Chemically pure endosulfan*: Colorless almost odorless crystals; αI mp 109°C, βI mp 213°C. Vapor pressure at 80°C: 9 × 10^{-3} mm Hg (pure active substance). Specific weight d^{20} 1.745 (pure active substance). Solubility: virtually insoluble in water, soluble in organic solvents (pure active substance).

Technical endosulfan with a content of > 94% active substance: Browny scale with possibly faint odor of sulfur dioxide; mp 80 to 90°C.

Stability: Stable in storage and stable against sunlight. By aqueous or alcoholic alkaline solutions and acids, endosulfan is hydrolyzed to endosulfandiol II.

5. Formulations of endosulfan.—Under the designations Thiodan®, cyclodan, thimul, thifor, malix, beosit, as well as mixed formulations with other insecticidal and acaricidal active substances, the following formulations are on the market:

5 EC

35 EC

35 WP: 50 WP

Dusts with 1 to 4% technical active substance

Granules with 1 to 5% technical active substance

25 ULV

c) Transformation products of endosulfan

Endosulfandiol (II) (endosulfan alcohol), $C_9H_8Cl_6O_2$, mol. wt. 360.9, mp 204°C (Frensch and Goebel 1954 a):

Endosulfan ether (III), $C_9H_6Cl_6O$, mol. wt. 342.8, mp 224°-226°C (decomp.) (Brace 1955):

Endosulfan sulfate (IV), $C_9H_6Cl_6O_4S$, mol. wt. 422.9, mp 181°C (Cassil and Drummond 1965):

Hydroxyendosulfan ether (V), $C_9H_6Cl_6O_2$, mol. wt. 358.9, mp 234°-236°C (Feichtinger and Linden 1958):

Endosulfan lactone (VI), $C_9H_4Cl_6O_2$, mol. wt. 356.8, mp 264.5°C (Riemschneider 1960):

As a further member in the chain of oxidized transformation products, Rückert and Ballschmiter (1972) could identify hydroxyendosulfan lactone (VII) by gas chromatography:

(VII)

Hydrophilic transformation products of endosulfan were mentioned by various authors, which products, however, could not be identified.

For the first time, Elzner (1973) was able to isolate, identify, and synthesize sulfuric acid esters of endosulfan transformation products from the urine of rats. The compounds possess the structural formulae VIII to X:

(VIII)   (IX)

(X)

d) Chemical and photochemical degradation of endosulfan and its metabolites

1. **Chemical degradation.**—Endosulfan (αI, βI) is hydrolyzed in alkaline solution to form endosulfan diol (II), from which one mol $Cl^{(-)}$/mol II is readily cleaved off; endosulfan lactone (VI) is likewise easily converted under evolution of 1.5 mol $Cl^{(-)}$/mol VI. Endosulfan ether (III) is scarcely attacked by aqueous alkali (Schuphan *et al.* 1972).

On boiling with alcoholic alkali, either of the isomers of endosulfan (αI, βI) is converted to a single compound, to which the structure XI was ascribed by Greve and Wit (1971), and the structure of the cyclic acetal XII by Hoch (1967 and 1972) as well as by Perscheid and Ballschmiter (1973):

XI   XII

Nucleophilic dechlorination at the double bond is the main course of degradation with such hexachloronorbornene derivatives only, that are substituted in 2- and/or 3-position by a hydroxymethylene group. Since in the case of derivatives without hydroxymethylene function such dechlorination only occurs sparingly, if at all, the special position of endosulfan among the other cyclodiene insecticides can be understood (Perscheid and Ballschmiter 1973, Schuphan *et al.* 1972).

Under ozonization in hexane, αI is but sparingly degraded, whereas approximately half of βI is so. In acetone/water, αI does not react with ozone, whereas the βI concentration drops significantly (Hoffmann and Eichelsdörfer 1971).

2. **Photochemical degradation.**—Endosulfan αI. βI, applied onto glass in a thin layer, yields after seven days of irradiation with UV light (*General Electric* germicidal lamp G 15 T 8) overwhelmingly endosulfan diol (II) and minor amounts of endosulfan ether (III), hydroxyendosulfan ether (V), endosulfan lactone (VI), and of an unknown compound. Irradiation of endosulfan diol (II) yields hydroxyendosulfan ether (V) and two other unknown compounds, which are also formed from hydroxyendosulfan ether (V), besides endosulfan ether (III). From endosulfan ether (III) there forms hydroxyendosulfan ether (V) and endosulfan lactone (VI). Irradiation of endosulfan lactone (VI) yields in less than 1% endosulfan diol (II) and the ether (III). The endosulfan sulfate (IV) stable under the applied photolysis conditions did not form during any of these photochemical reactions (Archer *et al.* 1972).

According to experiments by Geike (1970), the insecticidal activity of endosulfan is reduced by UV irradiation. The loss of activity is explained by the conversion of the active substances to more polar compounds whose penetration into the insect is hampered.

Schumacher *et al.* (1971) reported on the photochemical dechlorination of endosulfan by UV irradiation ($\lambda > 300$ nm). In hexane, αI is dechlorinated at the site of the nuclear double bond by reduction, whereas βI remains intact. In a dioxane/water mixture, reductive dechlorination of βI occurs at the methylene bridge.

Schuphan *et al.* (1972) and Schuphan and Ballschmiter (1972) reported on the photodegradation of endosulfan and its metabolites in methanol/water with borosilicate glass-filtered light ($\lambda > 300$ nm) and came to the following results: Either of the endosulfan isomers (αI, βI) as well as endosulfan diol (II) evolved 1 mol $Cl^{(-)}$/mol substance, whereas endosulfan sulfate (IV) and endosulfan lactone (VI) form in the same period of time 1.5 mol $Cl^{(-)}$/mol of compound. From αI and βI, two new compounds each are formed yet the dechlorination products from αI and βI differ from one another. The authors concluded therefrom that the sulfite ring remains intact, for if this were cleaved, identical compounds should form from αI and βI. All degradation products could be identified by gas chromatography, yet their structures could not be revealed.

According to experiments by Goebel (1971), endosulfan (αI and βI) as well as all its metabolites (II to VI) are completely degraded by irradiation with a mercury quartz dipping lamp (Hanau T Q 81) in water under aeration. Endosulfan (I) and endosulfan sulfate (IV) yield quantitatively Cl^-, $SO_4{}^{2-}$, and CO_2; endosulfan diol (II), endosulfan ether (III), hydroxyendosulfan ether (V), and endosulfan lactone (VI) yield quantitatively Cl^- and CO_2. On exclusion of air (under nitrogen) no degradation occurs. With wavelengths > 310 nm, only one to two Cl^-/mol are evolved from endosulfan and its metabolites.

Exposure to irradiation (cobalt 60, 5 Mrads) in hexane reduces the biological activity by 45% (Lippold *et al.* 1969).

Summary

Endosulfan, an insecticide and acaricide with a broad spectrum of activity, is prepared by diene synthesis of hexachlorocyclopentadiene and *cis*-butene-(2)-diol, or by hydrolysis of the Diels-Alder adduct from hexachlorocyclopentadiene and *cis*-1,4-diacetoxybutene-(2) to endosulfandiol and its subsequent conversion with thionyl chloride to the cyclic sulfite ester.

The technical active substance comprises two isomers, α- and β-endosulfan, which both possess similar biological activity. Endosulfan is virtually insoluble in water, and soluble in organic solvents. By acids and alkali, it is hydrolyzed to endosulfandiol.

The preparation is marketed in formulations as EC, WP, dust, granules, and ULV formulations with varying concentrations of active substance.

As transformation products and metabolites, there occur the following chemically and physically characterized compounds: endosulfan sulfate (IV), endosulfandiol (II), endosulfan ether (III), hydroxyendosulfan ether (V), endosulfan lactone (VI), as well as water-soluble conjugates of the aforesaid hydroxy compounds with sulfuric acid.

II. Analytical methods for endosulfan residues and metabolites
By S. Gorbach

a) Qualitative determination

A survey of the methods used until 1965 for identifying endosulfan (I) primarily by IR-spectroscopy and nuclear magnetic resonance spectroscopy has been presented by Maier-Bode (1968). A survey of the applications of nuclear magnetic resonance spectroscopy in pesticide analyses including endosulfan was published by Keith and Alford (1970).

Recently a few papers were published that are mainly concerned with the identification of peaks occurring in gas chromatograms. The problem to attribute beyond all doubt a peak to a specific substance is extremely difficult, if the chromatogram is rich in bands and there is not the slightest knowledge as to the nature of the components involved. Examples, therefore, are food control and control analyses in the scope of environmental protection.

A chemical modification of endosulfan resulting in a disappearance or a shifting of the peaks involved in a gas chromatogram was accomplished by Miller and Wells (1969) by the intercalation of an alkali-containing first column (25% KOH on 80 to 100 mesh Gas Chrom Q). Thus both of the peaks of the endosulfan isomers (αI, βI) are shifted toward shorter retention times. The authors present retention times of 32 of the most employed pesticides, both before and after chemical modifications. After the reaction with alkali, the following active substances are no longer eluted: α-BHC, diazinon, lindane, methyl parathion, Ronnel, heptachlor, malathion, parathion, heptachlor epoxide, ovex, perthane, tetradifon, and carbophenothion. Derivatization with methanolic or ethanolic potassium hydroxide solution was carried out by Greve and Wit (1971).

The reaction with ethanolic potassium hydroxide solution was made part of a multi-residue method by Young and Burke (1972). It is especially gratifying that it was possible to separate and differentiate—besides endosulfan (I)—a series of other insecticides, *e.g.*, DDT, from the polychlorinated biphenyls, since the latter are stable to alkali in the used concentration and do not show any change of their GC fingerprint.

The reduction of endosulfan (I) with lithium aluminum hydride in tetrahydrofuran and subsequent silylation of the thus formed endosulfandiol (II) was described by Chau (1969). Alternatively, the endosulfandiol (II) may also be esterified to the corresponding diacetate. The author discusses the IR spectra as well as the NMR spectra of the silyl—and acetyl derivatives of endosulfandiol. The retention times in gas chromatography before and after derivatization are given. In a later paper, Chau and Terry (1972) use an aluminum oxide microcolumn charged with acetic acid anhydride and sulfuric acid for direct re-esterification of endosulfan (I) to endosulfan diacetate. The solution is passed onto the column, washed in, and caused to react at $100°C$ (2 hr). A comparison of this method with that of Greve and Wit (1971) revealed better signals with the electron capture detector. The reason therefore lies in the relative insensitiveness of the electron capture detector for the derivatives (XI or XII) formed in the reaction with alkali. Figure 1 shows chromatograms before and after derivatization of endosulfan to endosulfan diacetate after Chau and Terry (1972). A simple method for confirmation of endosulfan residues by conversion with hydrochloric acetic anhydride is described by Musial *et al.* (1976). In two further papers, Chau (1972) and Chau and Terry (1974), endosulfan (I) was quantitatively converted to endosulfan ether (III) on aluminum oxide columns charged with sulfuric acid at 100°C (reaction time 15 hr), and the reaction product was gas-chromatographically determined. Conversion to the diol (II) succeeds after Chau and Lanouette (1972) on aluminum oxide columns charged with KOH or potassium-*tert.*-butyl alcoholate.

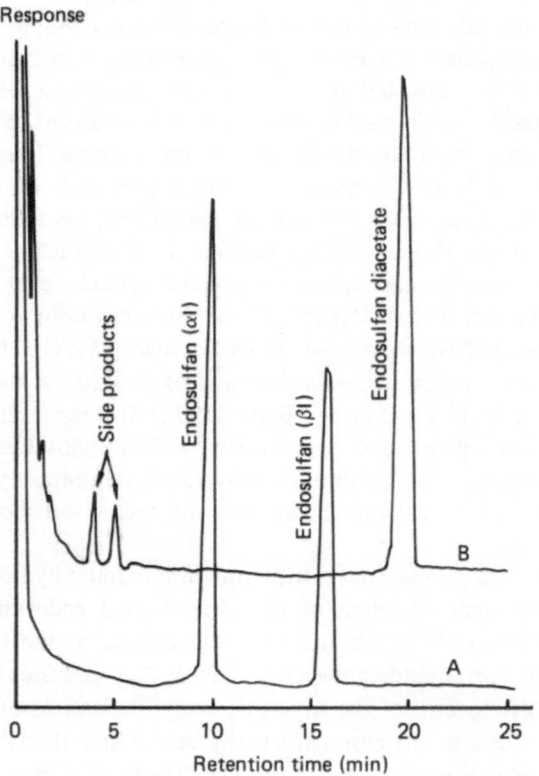

Fig. 1. Chromatograms of 100 pg. each of α-endosulfan and β-endosulfan before (A) and after (B) acetylation in solid matrix at 100°C for 2 hr. (Chau and Terry 1972).

b) Quantitative determination

1. In formulated products and in the technical grade substance.—Maier-Bode (1968) has presented a compilation of the methods that have become known until 1967. The CIPAC method already mentioned by Maier-Bode has meanwhile appeared in CIPAC Handbook 1 (1970), *W. Heffner and Sons, Ltd., Cambridge, England.* For the analysis of the technical grade substance the reaction of endosulfan (I) with iodine and retitration of the unconsumed iodine has proved good (Graham *et al.* 1964). More recently gas chromatographic determination methods for the active substance and some possible impurities such as endosulfandiol (II) and endosulfan sulfate (IV) were worked out as firm-employed methods (*Hoechst* 1972, *Niagara* 1971). *Hoechst* (1972) uses for separation 10% DC-LSX-3-0295 on Chromosorb W-AW-DMCS, 2 m and *Niagara* (1971) 25% QF-1 on Gas Chrom, 4 ft.

2. Residue analysis.—

α. *Extraction.*—Mills *et al.* (1963) introduced acetontrile as extractant in residue analysis.

Acetonitrile has also proved successful for the extraction of endosulfan (I) and its most prominent metabolite endosulfan sulfate (IV), especially so, because the purification step by shaking out with hexane may follow immediately [for tobacco Hengy and Thirion (1971), for plant material Diemair *et al.* (1968), Kadoum (1969), and Thier (1972)]. Burke and Porter (1967) have found though, that the extraction efficiency decreases should the water content of the sample drop to below 75%. Burke *et al.* (1971) have confirmed this observation in the extraction of endosulfan residues and recommended the mixture of acetonitrile with 35% water proposed by Bertuzzi *et al.* (1967). Root-absorbed residues, however, resist even such extractants and must be extracted with chloroform/methanol (1/1) in a Soxhlet apparatus (exhaustive extraction by Wheeler *et al.* 1967). The dependence of extraction efficiency on the water content of the sample is shown in Figure 2.

An interesting extraction procedure for fats and oils was described by Porter and Burke (1973). The sample is mixed with Florisil, filled into a chromatography tube, and the active substances including endosulfan are then eluted with acetonitrile (plus 10% water).

For the extraction of plant protective agents, including endosulfan, from blood, Griffith and Blanke (1974) recommended a mixture of hexane and acetone (9 + 1) after addition of sulfuric acid (2.5 ml 60% H_2SO_4/ml blood).

Exhaustive extraction in a Soxhlet with chloroform/methanol (9/1) is also used by Dorough and Gibson (1972) for the extraction of endosulfan from tobacco and by Schlunegger (1968) for the extraction from organ material.

Ten % methanol in dichloromethane was used by Smith *et al.* (1972) for Soxhlet extraction of total diet samples.

The formerly much used benzene in mixture with, *e.g.*, alcohols such as isopropanol (Maier-Bode 1968), is used further in altered mixing ratios and

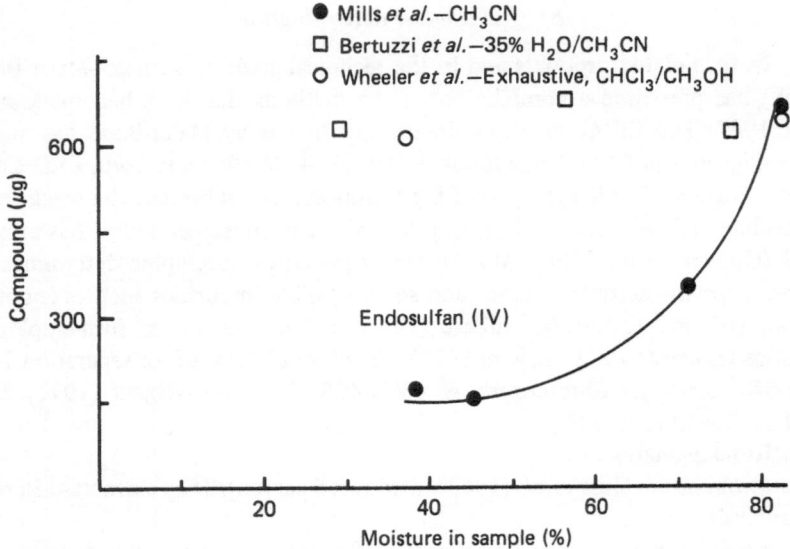

Fig. 2. Endosulfan (isomers I and II) and endosulfate residues from foliar application; extraction vs. dehydration of sample (Burke *et al.* 1971).

with different partners of mixture for the extraction of endosulfan (I) and endosulfan sulfate (IV). Thus, *e.g.*, McLeod and Wales (1972) used benzene/acetone
(19/1) for the extraction of biological material in a Soxhlet, and Martens (1972)
used benzene/methanol for the extraction of fungus mycelium. For routine
analyses, however, the toxic benzene should not be used.

Hexane, alone and in mixture with acetone, is also a frequently employed
extractant. Himel and Uk (1972) as well as Uk and Himel (1972) extracted
insects therewith, El Zorgani and Omer (1973) fungus mycelium in a Soxhlet,
and Keil *et al.* (1972) tobacco.

Beard and Ware (1969) used a mixture of hexane, ethanol, and petroleum
ether (2/1/1) for the extraction of plant material, and Sissons *et al.* (1968) a
mixture of hexane and acetone (1/4) for cold-extraction of vegetables.

The last-mentioned authors critically commented on the extraction procedures used until 1968, and they claimed the use of a hexane-acetone mixture
to be as good as an exhaustive extraction in a Soxhlet. In a second report (Sissons and Telling 1970), this extraction procedure is described, incorporated into
a multi-method. A mixture of diethylether and hexane (3 + 97) was used by Levi
et al. (1972). The Pesticide Analytical Manual, *U.S. Department of Health, Education and Welfare, Food and Drug Administration,* Edition 1972, Section 211,
gives further detailed process instructions within the scope of a multi-method.

In the multi-method for organochlorine and organophosphorus insecticides
of *Deutsche Forschungsgemeinschaft* (1968/1972) Methode S 5, fruits and vege-

tables are extracted with acetone, cereals with methanol/chloroform, and fats with petroleum ether. Acetone is also the preferred extractant in the multi-method of Becker (1971).

For the extraction of endosulfan (I) from air, Tessari and Spencer (1971) use a Nylon chiffon cloth screen previously soaked in glycol and exposed for 5 days methane. Gorbach *et al.* (1971 a and b), Knauf *et al.* (1976), and Bauer (1972) use hexane for extracting from water.

For the extraction of endosulfan (I) from air, Tessari and Spencer (1971) use a nylon chiffon cloth screen previously soaked in glycol and exposed for 5 days to the contaminated air.

β. Cleanup.—A survey of the methods described until 1967 was given by Maier-Bode (1968).

In the later development, partition procedures followed, as a rule, by column chromatography, are found as a first step in the cleanup of extracts. In many cases, column chromatography is replaced by adsorption of impurities to absorbents in the batch procedure.

The standard methods of *Pesticide Analytical Manual* (1971) as well as *Methodensammlung der Deutschen Forschungsgemeinschaft* (1968/1972) prescribe for extracts from fatty material (butter, etc.) the partition between acetonitrile and petroleum ether, or hexane, respectively. Differing therefrom, Bro-Rasmussen *et al.* (1968) and Schlunegger (1968) used the distribution between *n*-hexane and dimethylformamide proposed by Faubert Maunder *et al.* (1964).

In practically all cases, the acetonitrile phase containing the endosulfan after partition with petroleum ether (hexane), is diluted with water, and the endosulfan is then extracted with a suitable solvent (*Pesticide Analytical Manual* 1972, *Methodensammlung der Deutschen Forschungsgemeinschaft*, Methode 50 and S 5c, 1968/1972).

Should extraction be performed directly with acetonitrile, and should lipids be present in but negligible amounts, partition with hexane does not apply, and the extraction of the aqueous acetonitrile follows immediately.

Some authors extracted with a mixture of polar/nonpolar solvents and washed the polar solvent out in a cleanup step with water, *e.g.*, hexane/petroleum ether/ethanol (Beard and Ware 1969), acetone/hexane (Sissons *et al.* 1968). The previously, but in a different connection, often-mentioned purification by freezing out of primarily lipid-like impurities, is again recommended by McLeod and Wales (1972) within a multi-method for endosulfan (I), too. After accomplished extraction, the extractant acetone/benzene (19/1) is chilled to −78°C and then cold-filtered after 30 min. The solution obtained can be immediately analyzed gas-chromatographically.

In most cases the cleanup with acetonitrile is followed by a further column chromatographic step.

Florisil® is used as separation phase and mentioned as cleanup standard (*Pesticide Analytical Manual*, 1972) also for the separation of endosulfan (I). Dorough and Gibson (1972) employed Florisil for the cleanup of tobacco extracts, and Schlunegger (1968) found Florisil in mixture with active carbon to be suitable for the cleanup of extracts from organ tissues. DDT is eluted with a mixture of n-hexane and 4% toluene and subsequently endosulfan (αI and βI) is eluted with a mixture of equal proportions of acetone and diethylether. Florisil, as stationary phase in separation columns, was further used by Göke (1972), Tessari and Spencer (1971), Levi et al. (1972), and Thier (1972). Hengy and Thirion (1971) likewise used Florisil for residue determination of endosulfan in tobacco. DDT is first eluted with hexane, and endosulfan (αI and βI) and endosulfan sulfate (IV) are subsequently eluted with benzene.

Besides Florisil, aluminum oxide proved very suitable for column chromatographic separation of both endosulfan (αI and βI) and endosulfan sulfate (IV). *Hoechst* (1972) uses aluminum oxide of activity grade III. Endosulfan (αI and βI) as well as endosulfan sulfate (IV) are eluted with 250 ml of benzene. Bro-Rasmussen et al. (1968) used aluminum oxide with 5% water and eluted with petroleum ether. Diemair et al. (1968) used acid aluminum oxide, activity grade II, and eluted with petroleum ether/diethylether (7/3), and Sissons and Telling (1970) used hexane and acetone for elution.

Kadoum (1968) suggests silica gel as separating material for cleanup of complex insecticide mixtures. Without endeavouring to separate the insecticides present, the author elutes 23 different agents quantitatively with benzene. Maier-Bode (1968), in his compilation, already referred to an effective cleanup of extracts by means of Nuchar-Attaclay. Meanwhile further authors have successfully used said carbon mixture in the residue analysis of endosulfan (I) (Tappan et al. 1967, *Hoechst AG* 1968, Beard and Ware 1969, Göke 1972, Uk and Himel 1972). Sissons et al. (1968) recommend substituting the less toxic toluene as solvent and suspending agent for the otherwise conventional benzene. Becker (1971) uses instead a silica gel-carbon mixture as cleanup substance in a multi-method. Quite recently the cleanup by gel permeation chromatography became more important. This procedure especially lends itself for the treatment of fat-containing material and was applied, e.g., by van Dyk et al. (1978) in the analysis of chicken eggs. In one case a sublimation procedure was successfully carried out for the isolation of endosulfan (I) from hay samples (Schmidlin-Mészáros and Romann 1971).

γ. *Gas chromatographic determination.*–In residue analysis, preferably gas chromatography is used for the determination of endosulfan (I).

Summarized descriptions of the separation columns and detectors used have been presented by Maier-Bode (1968) with emphasis on endosulfan (I), and Ebing (1970, 1973, and 1974) in a bibliography on gas chromatography of plant-protective agents. In these surveys, there are numerous references to papers, including detailed working specifications.

As was already stressed in the summary of Maier-Bode (1968), the pre-treatment of the carrier of the stationary phase appears to be decisive for the stability of the active ingredient during its passage through the column. Whereas in earlier years this pre-treatment had to be carried out partly by the analyst himself, carriers are now commercially available which are already acid-washed and pre-treated with dimethyl dichlorosilane, such as Chromosorb-W-AW-DMCS, or Chromosorb-G-AW-DMCS, respectively. As literature citations exemplary for many others: Archer et al. (1972), and Chau and Wilkinson (1972).

Moreover, carriers such as Gas Chrom Q (Eichelberger and Lichtenberg 1971), Chromoport (Hengy and Thirion 1971), and Anachrom-ABS (Dorough and Gibson 1972) were cited, for which one reference each may stand for many others.

Stationary phases that are used for separating endosulfan (I) from other substances are more numerous than are carriers. They are all based on silicon, yet of different polarities, and they are used in increasing measure in various ratios of mixture in order to solve specific separation problems.

A survey of separation phases frequently employed for endosulfan (I) so far is given in Ebing (1970, 1973, and 1974).

The development of a separation column that may be operated at high temperatures (up to 400°C) is of interest (Bowman and Beroza (1971). Impurities are mainly held back in the column inlet by means of a frequently exchangeable glass wadding stopper. The column itself is cleaned by 1-hr elution at 400°C. The authors give the retention times of 146 active substances, including endosulfan (I).

For detection, an electron capture detector is mostly used, of which mainly two types are now available, provided with varying ionization sources (tritium or [63]nickel).

The detection of endosulfan (I) with the detector fitted with tritium has been sufficiently described and concisely presented by Maier-Bode (1968). Even nowadays this detector is readily employed for residue analyses (Dorough and Gibson 1972, El Zorgani and Omer 1973, Chau 1972, Abbott et al. 1969, Beard and Ware 1969, Chau and Terry 1972, etc.). The extended studies dealing with residues in food in the U.S.A. (Corneliussen 1969, 1970, and 1972) were also carried out with this type of detector.

The detector equipped with [63]Ni as ionization source increasingly gains importance. One advantage of this detector lies in its higher thermal resistance (300°C) which allows of operating the detector at higher temperatures resulting in a lower risk as to precipitation of impurities. Hence this detector is frequently encountered in recent publications (e.g., Rückert and Ballschmiter 1972, Eichelberger and Lichtenberg 1971, Uk and Himel 1972). The much used microcoulometer of the past—Maier-Bode has given a summary—is but seldom mentioned in recent literature for the determination of endosulfan (I). As an alternative determination for the confirmation of findings with the electron-capture detector and the determination procedure for endosulfan (I), the micro-coulometer is referred to in the "Pesticides Analytical Manual" (1972) and in "Methodensammlung der Deutschen Forschungsgemeinschaft" (1968/1972).

Similarly as an auxiliary for identity confirmation, the flame photometric detector is used by its employment in the sulfur-sensitive form (sulfur mode), e.g., by Keil et al. (1972) for confirming the evidence of endosulfan (I) in tobacco. For detection, the light emission (393 nm) is made use of that occurs when sulfur-containing compounds are conducted into an oxyhydrogen gas flame.

In connection with endosulfan (I), Bowman and Beroza (1969) describe another flame photometric detector in which the intensity of flame coloration is measured, that occurs when chlorine-containing compounds are passed into a flame in the presence of copper.

Other determination methods such as photometric and total chlorine determinations, as they were still described in the survey by Maier-Bode (1968), have not found access in recent publications.

However, semiquantitative methods for estimating the order of magnitude of endosulfan (I) residues following thin-layer chromatographic separation are still to be found. Voigt and Noske (1968), e.g., make endosulfan visible with thymol, and they compare the color depth of spots of known amounts with the color depth of the spots resulting from the samples.

δ. *Thin-layer chromatographic separation.*—Especially in the quantitative determination of endosulfan (I) the thin-layer chromatography plays a minor role. It is currently used as an auxiliary means for identification and a rather qualitative general view, in particular for studying degradation processes. Belliveau et al. (1970) make visible sulfur-containing active substances, including endosulfan (I), by adding bromine to the mobile phase and subsequently reexposing the plate to bromine vapors, which step is followed by spraying the plate first with an iron-III-chloride or zinc chloride solution, and then with a chelate former, e.g., 8-hydroxyquinoline. The spots become visible under UV light, and even 0.1 to 0.3 μg of substance can be evidenced.

Geike (1970 a and b), in a series of communications, deals with the enzymatic identification of numerous insecticides, including endosulfan (I). In his first communication (Geike 1969), the active substances separated on the thin-layer plate are sprayed consecutively with β-naphthyl acetate, Genuine Blue Salt B, and esterase from neat's liver (pH 7). All tested insecticides, i.e., endosulfan, too, yet exclusive of DDT, DDE, DDD, perthane, and methoxychlor, shown an inhibitory effect to the enzyme.

In 1970 a publication followed in which identification was performed with trypsin (Geike 1970). The plate was first sprayed with the enzyme solution, then with sodium benzoyl-DL-ariginin-4-nitro-anilide hydrochloride. Endosulfan (I) inhibits the enzyme under the given test conditions. In 1971 (Gieke 1971 a and b) two further papers followed with amylase and phosphatase as enzymes. The developed plates were first sprayed with α-amylase from *Bacterium subtilis* or β-amylase from rye, and after 30 min at 25°C with a starch solution. Endosulfan (I) inhibits the amylase, evidenced by a blue spot on a bright background. The treatment of the plates with alkaline phosphatase and naphthyl phosphate as substrate, followed by application of Genuine Blue Salt B as diazo component, gave a good test for endosulfan (I) (2 to 150 μg).

Ebing (1969) described a technique for obtaining improved reproducibility of R_f values in thin-layer chromatography. The thin layer was sandwiched between two glass plates and developed after prior application of the sample material. Silica gel G and polyamide serve as separation gel. Endosulfan (I) is among the 17 examples given in said paper.

Further indications of suitable thin-layer chromatographic separation possibilities are found in Maier-Bode (1968), Schmidlin-Mészáros and Romann (1971), and Voigt and Noske (1968). The *U.S. Food and Drug Administration*, in "Pesticides Analytical Manual" (1972), Section 410, recommends thin-layer plates carrying aluminum oxide as confirmatory test for 44 plant-protective agents or their transformation products, including endosulfan. As mobile phases there are given:

(a)	n-heptane	R_{TDE}	endosulfan I	0.88
			endosulfan II	0.07
(b)	2% acetone in n-heptane	R_{TDE}	endosulfan I	0.92
			endosulfan II	0.24
(c)	25% dimethylformamide in diethylether	R_{TDE}	endosulfan I	3.1
			endosulfan II	0.0

R_{TDE} = relative value of migration distance of endosulfan to that of p,p'-TDE [2,2-bis(p-chlorophenyl)-1,1-dichloroethane].

ϵ. *High-performance liquid chromatography as determination method for endosulfan.*—A "reversed phase" liquid chromatographic determination is described by Demeter and Heyndrickx (1976). The authors used as stationary phase a chemically bound C_8 and C_{18} RPLC column and mixtures of water and methanol as mobile phases. The most important transformation products, *e.g.*, endosulfandiol (II), endosulfan ether (III), endosulfan lactone (VI), and endosulfan sulfate (IV) are included in the investigations.

Besides this more fundamental paper, Hoodless *et al.* (1978) described the high-performance liquid chromatography for the analysis of agents for the treatment of plants that are contained in "European Economic Community Directive on Fruit and Vegetables" (including endosulfan). As separation column, 5 μm ODS-bound silica gel (Spherisorb ODS) is used, and as mobile phase mixtures of acetonitrile/water. There were investigated 41 substances (active substance and metabolites).

ζ. *Recommended methods for the determination of endosulfan residues.*—A tremendous number of market basket analyses, including the determination of endosulfan, is currently carried out in the U.S.A. (Corneliussen 1970 and 1972). All investigations are operated according to the procedures described in "Pesticide Analytical Manual" of the *U.S. Food and Drug Administration, U.S. Department of Health, Education and Welfare,* Vols. I and II. The "Pesticide Analytical Manual" exists in several revised editions. The last is dated September 1972. According to it, the extraction of nonfatty material is first performed with aqueous acetonitrile. Subsequently, the acetonitrile extraction filter cake is subjected to an exhaustive extraction with chloroform/methanol (1/1) in a Soxhlet. Fatty material is directly extracted in the Soxhlet with the chloroform/methanol mixture. The methanol is washed out of the extract with water, and the aqueous solution is re-extracted with petroleum ether. Chloroform is evaporated in the presence of isooctane and united with the petroleum ether extract of the aqueous solution.

Only if it is proven by comparative analysis that nonfatty material is quantitatively extractable with aqueous acetonitrile, and fatty material is so according to one of the methods described under Section 211.13, the Soxhlet extraction may be dropped, a decisive factor in case of extensive series of analyses involved. Should the extract hold a major amount of fat, it is dissolved in petroleum ether and shaken out with acetonitrile (Section 211.14 a). Then follows the column-chromatographic separation on Florisil (Section 211.14 d). Elution with 15% diethyl ether in petroleum ether elutes endosulfan αI and partly endosulfan βI. The remaining endosulfan βI is eluted with 30% diethyl ether in petroleum ether, and endosulfan sulfate (IV) with 50% diethyl ether in petroleum ether.

The determination of endosulfan (I) is gas-chromatographically operated with the electron capture detector, and the identity confirmation with the micro-coulometer in the sulfur-sensitive form ("Pesticide Analytical Manual" 1972, Vol. I, Section 301). As separator column there is used: 1.8 m (6 ft) × 4 mm id. glass column packed with 15% QF1 + 10% DC 200 on Gas Chrom Q. Relative to aldrin, endosulfan αI, endosulfan βI, and endosulfan sulfate (IV) possess the retention times 1.89, 2.92, and 5.40.

Blass (*Deutsche Forschungsgemeinschaft,* "Methodenbuch," 3rd Ed. 1974) presents a further recommended multi-method. Heavily fat-containing samples are dissolved in petroleum ether and shaken out with acetonitrile. Fruits and vegetables are extracted with acetone, the extract is diluted with water and shaken out with dichloromethane. Cereals are mixed with Celite, filled into a column, and extracted with chloroform/methanol.

In each case the extracts are cleaned up by chromatography on aluminum oxide. For cereals, an additional cleanup is carried out by chromatography on Florisil.

Gas chromatographic determination is performed on three gas chromatographic separator columns with varying polarity:

Column (a) 1% DC 200, Chromosorb W, glass, length = 1.50 m, 2.8 mm id.
       (b) 2% silicon rubber (*Merck*), Chromosorb G, acidwashed, silanized, glass, length = 1.50 m, 2.8 mm id.
       (c) 15% QF 1 and 5% silicon fluid 550 (*Merck*) on Chromosorb W, acidwashed, glass, length = 1.50 m, 2.8 mm id.

The retention times for endosulfan (I) are not given, but a recovery rate of 77 to 82% is communicated for endosulfan (I).

Methods for the determination of endosulfan (I) without considering other plant-protective agents are compiled in "Pesticide Analytical Manual," Vol. II (1972), method 50.

c) Analysis of endosulfan metabolites

1. **Introduction.**—Maier-Bode (1968) gives a survey of the methods known until 1968 for the separation, identification, and determination of endosulfan degradation products.

Of the metabolites that have become known so far, only endosulfan sulfate (IV) is toxicologically of importance. Hence the analytical procedures described for endosulfan (I) itself also comprise, as a rule, endosulfan sulfate (IV). This especially applies for the methods described in "Methodenbuch" der *Deutschen Forschungsgemeinschaft* 1972 and in "Pesticide Analytical Manual" of 1972. In the course of the numerous papers on the behavior of endosulfan (I) in the environment, quite a series of analytical procedures have been developed that allow of extracting, separating from one another, identifying, and/or determining besides endosulfan sulfate (IV) also endosulfan diol (II), endosulfan ether (III), and hydroxyendosulfan lactone (VII) from organic material.

2. **Extraction.**—Archer (1973) extracts in a Soxhlet with benzene/propanol (95/5) from plants the endosulfan isomers (I) and the metabolites endosulfan diol (II), endosulfan ether (III), hydroxyendosulfan ether (V), hydroxyendosulfan lactone (VII), and endosulfan sulfate (IV). Beard and Ware (1969) use for extraction of endosulfan sulfate (IV), endosulfan diol (II), and endosulfan ether (III) a mixture of hexane/ethanol and petroleum ether and shake out the ethanol with water. Endosulfan (I) and its degradation products remain in the organic phase.

Schmidlin-Mészáros and Romann (1971) extract endosulfan (I) together with endosulfan sulfate (IV), endosulfan diol (II), and endosulfan ether (III) by using benzene/isopropanol.

From microorganisms, endosulfan diol (II) can be extracted with acetone (*Hoechst AG* 1971), benzene/methanol (1/1) (Martens 1972), and toluene/methanol (Perscheid *et al.* 1973).

In animal products such as meat, fat, or milk, no metabolites except endosulfan sulfate (IV) have been found so far. As to endosulfan sulfate (IV), the same extraction methods as already described for the parent substance itself have proved successful.

For the isolation of endosulfan diol (II), Schlunegger (1968) indicates a procedure. In the urine of endosulfan-treated animals hydrophobic metabolites are found in a minor amount, whereas hydrophilic metabolites occur in major amounts. The lipophilic proportion is readily extracted with benzene (Gorbach *et al.* 1968, Schuphan *et al.* 1968, Elzner 1973), whereas the hydrophilic proportion is extractable with benzene to a part after hydrolysis of the conjugates only

(Elzner 1973). The isolation of the total hydrophilic metabolites is feasible by thin-layer chromatography (Elzner 1973).

3. **Cleanup.**—Archer (1973) cleans the endosulfan metabolites (II, III, IV, V, and VI) up by column chromatography on activated Florisil and elutes with a mixture of propanol/diethyl ether/pentane (2/28/70). Beard and Ware (1969) describe the cleanup of extracts with Nuchar-Attaclay and record for the metabolites IV, II, and III recoveries of over 90%.

4. **Gas-chromatographic determination.**—Information on the gas-chromatographic separation and determination of endosulfan metabolites is contained in the publication by Maier-Bode (1968). Schuphan *et al.* (1968) (cp. Table I) describe in detail stationary phases suited for the separation of endosulfan metabolites.

Retention times of the silylated and acetylated endosulfan diol (II) are also found in Chau (1969). For separation, the author uses a mixed column (4% DC 11 and 6% QF1 (1/1) on Chromosorb W), and for detection an electron capture detector. In a later paper (Chau and Terry 1972) a stationary mixed phase consisting of 3.6% OV-101 and 5.5% OV-210 is used for the separation of endosulfan diol diacetate (XIII). For separating endosulfan diol (II), endosulfan sulfate (IV), and endosulfan ether (III), Beard and Ware (1969) use a 2-ft × 1/4 in od column with 5% SF 96 on Chromosorb W.

Table I. *Retention values of endosulfan and its transformation products* (Schuphan *et al.* 1968).

Substance	Abbrev.	XE-60	QF-1	DC-200	SE-52	XE-60:SE-52 (1:1)
α-Endosulfan (I)	αE	0.37	0.68	1.45	1.30	0.65
β-Endosulfan (I)	βE	1.10	1.20	1.95	1.90	1.36
Endosulfan sulfate (IV)	ES	2.75	2.80	2.50	2.50	2.60
Endosulfan-diol (II)	ED	0.23	0.31	1.95	1.05	0.46
Endosulfan ether (III)	EE	0.17	0.27	0.60	0.52	0.28
α-Hydroxy endosulfan ether (V)	HE	0.42	0.35	0.95	0.95	0.49
Endosulfan lactone (VI)	EL	1.00	1.00	1.00	1.00	1.00
	M_1		2.60			
	M_2		0.74			

Columns header: Relative retention times[a]

[a] 2% stationary phase on Anakrom ABS, 110-120 mesh. Glass columns id 2.5 mm; length 1 m; column temp. 195°C. Gas chromatograph: Varian-Aerograph 205 B with two electron capture detectors. Carrier gas: nitrogen, re-purified. Retention times of (EL): XE-60 8.0 min; QF-1 5.2 min; DC-200 6.5 min; SE-52 1.85 min; XE-60:SE-62 1:1 3.7 min.

Gorbach *et al.* (1968) use 10% QF1 and alternatively 10% SE 30 for separating endosulfan diol (II) and hydroxyendosulfan ether (V).

Archer *et al.* (1972 and 1973) separate the endosulfan metabolites endosulfan sulfate (V), endosulfan diol (II), hydroxyendosulfan ether (V), and endosulfan lactone (VI) on 5% DC 710 silicon fluid + 5% SE 30 silicon rubber on Chromosorb W-AW-DMCS (length 2.4 m, id = 3 mm) and employ the electron capture detector for detection.

In the first paper (1972), also, the retention times of the acetylated metabolites are tabulated. El Zorgani and Omer (1973) separate and determine endosulfan sulfate (IV) and endosulfan diol (II) on 2.5% SE 52 + 0.5% Epikote on Chromosorb W-DMCS.

In a later study, Perscheid *et al.* (1973) likewise use XE 60 as stationary phase, yet they additionally introduce OV 225 as a suitable stationary phase (cp. Table II).

Rückert and Ballschmiter (1972), during their investigation on the degradation of endosulfan (I), produced methyl and ethyl derivatives especially of the cyclic hydroxylactone and separated them from one another by gas chromatography (cp. Table III).

5. Thin-layer chromatographic separation and identification.—Archer (1973) separates the metabolites II to VI from one another and from the two endosulfan isomers αI and βI on silica gel with acetone/hexane (20/80) and makes visible the single components with the 2-phenoxyethanol spray reagent of Mitchell (1958). The R_f values under these separation conditions are shown in Table IV.

Schuphan *et al.* (1968) separate the metabolites initially referred to (I to IV) on a mixed silica gel polyamide phase and believe the benzene/ethanol (9 + 1) mixture to be the best mobile phase for separation. For the separation of endosulfan diol (II) and hydroxyendosulfan ether (V), Gorbach *et al.* (1968) use acetone/hexane (1/1) as mobile phase and Rhodamin B as spray reagent.

Table II. *Relative retention times of various endosulfan metabolites* (Perscheid *et al.* 1973).

Compound	Stationary phase 2 %[a]	
	XE 60	OV 225
Endosulfan α (I)	0.48	0.45
Endosulfan β (I)	1.22	1.21
Endosulfan sulfate (IV)	2.59	2.57
Endosulfan-diol (II)	0.34	0.24
Hydroxyendosulfan ether (V)	0.68	0.59
Endosulfan lactone (VI)	1.00	1.00
Endosulfan ether (III)	0.22	0.19

[a] Column: length = 1.0 m, id = 2.5 mm; carrier: Gas-Chrom Q.; thermostat temperature: 190°C; Electron capture detector (tritium).

Table III. *Relative retention of some endosulfan metabolites and derivatives* (Rückert and Ballschmiter 1972).

Substance	Relative retention times, stationary phase[a]		
	OV 225	QF 1	XE 60
R == H	5.56	3.19	6.4
R = OCH$_3$	2.28	1.94	2.3
R = OC$_2$H$_5$	2.58	2.50	2.55
(structure)	7.89	–	–
Aldrin	1.0	1.0	1.0

[a] Column: 1 = 1 m, id = 2.5 mm; carrier: Anachrom ABS; thermostat temperature 200°C; electron capture detector (^{63}Ni).

6. High-performance liquid chromatographic separation.—A high-performance liquid chromatographic separation of the metabolites endosulfan diol (II), endosulfan ether (III), endosulfan sulfate (IV), and endosulfan lactone (V) was described by Demeter and Heyndrickx (1976).

Summary

The report covers the analysis procedures dealt with in the literature for the determination of endosulfan (I) and its metabolites. For presenting a better survey, first the possibilities for identification of endosulfan (I) are discussed. Then

Table IV. *Retention times and R_f values of endosulfan and its degradation products* (Archer 1973).

Compound	Retention time (min)	R_f value[a]
Endosulfan α (I)	4.5	0.60
Endosulfan β (I)	7.4	0.35
Endosulfan diol (II)	3.8/8.5	0.20
Endosulfan sulfate (IV)	10	0.16
Endosulfan ether (III)	1.8	0.45
Hydroxyendosulfan ether (V)	2.5/3.0	0.20
Endosulfan lactone (VI)	4.0	0.22

[a] 5% DC 710 + 5% SE 30 on Chromsorb A-AW-DMCS (2.4 m, 3 mm id).

the quantitative determination of the active substance and its formulations is treated. This is followed by a section in which the procedures for determination of endosulfan (I) residues are discussed. Said section is subdivided, dealing with processes for the extraction of residues from different substrates, the cleanup of the extracts, and the determination of the isolated residues. A special paragraph is dedicated to thin-layer chromatographic separation and identification of endosulfan (I). In order to give some guidance to the practitioner, a paragraph describes a few analysis procedures recommendable on account of their application and recognition in practice.

Finally the processes for the determination of endosulfan (I) metabolites are discussed. As was the case with endosulfan (I), the description is likewise subdivided in dealing with extraction, cleanup, and determination procedures.

III. Metabolism of endosulfan in plants and animals

By S. Gorbach

a) Plants

Maier-Bode (1968) gives a survey of the endosulfan behavior on plants. It follows therefrom that endosulfan is not to be characterized as systemic, and that it metabolizes on the plant surface essentially to endosulfan sulfate (IV). In contrast thereto, Beard and Ware (1969) have found a minor translocation of foliar-applied endosulfan and endosulfan derivatives. The order of translocation decreases in the sequence endosulfan β (I) > endosulfan sulfate (IV) > endosulfan ether (III) > endosulfan α (I) > endosulfan diol (II). These authors, too, could not detect, besides endosulfan sulfate (IV), any other metabolites, not even endosulfan diol (II).

An interesting paper on the behavior of endosulfan (I) residues during drying of lucerne in the dark under UV or sunlight was presented by Archer (1973).

Quite in contrast to the findings recorded to date, especially to those by Harrison *et al.* (1967), Archer (1973) comes to the result that the endosulfan sulfate (IV) preferably forms in the dark during the drying (of lucerne) (cp. Fig. 3).

Chopra and Mahfouz (1977) reported on an extensive study concerning conversion of endosulfan in tobacco leaves. It is remarkable that the authors, among other facts, observed a re-conversion of endosulfan sulfate (IV) to endosulfan (αI). A survey of the degradation and conversion processes is shown in Figure 4.

b) Microorganisms (fungi, bacteria, algae)

Gorbach and Knauf (1971) have observed that ^{14}C-endosulfan (βI) is readily taken up by aquatic microorganisms and metabolized to endosulfan diol (IIA). In corresponding tests with sewage sludge, out of 30 mg endosulfan (IA)/1.5 g bacterial dry mass, 50% of the used endosulfan was converted within 7 days. Similar to bacteria, also green algae (*Chlorella, Scenedesmus*) are able to degrade endosulfan (IA) to endosulfan diol (IIA).

Martens (1972, 1976, and 1977) extensively studied the degradation of ^{14}C-labeled endosulfan (IA) by 28 soil fungi strains and 50 bacteria strains in shaking cultures for 10 days.

Half of all investigated fungi forms endosulfan sulfate (IV). Foremost lies the strain E54 *Penicillium* sp. with 75% conversion of the given endosulfan (IA), followed by E20 *Aspergillus* sp. with 66% and E9 *Botrytis cinerea* with 64% conversion. Endosulfan diol (IIA) is formed by only a few strains and then but in a

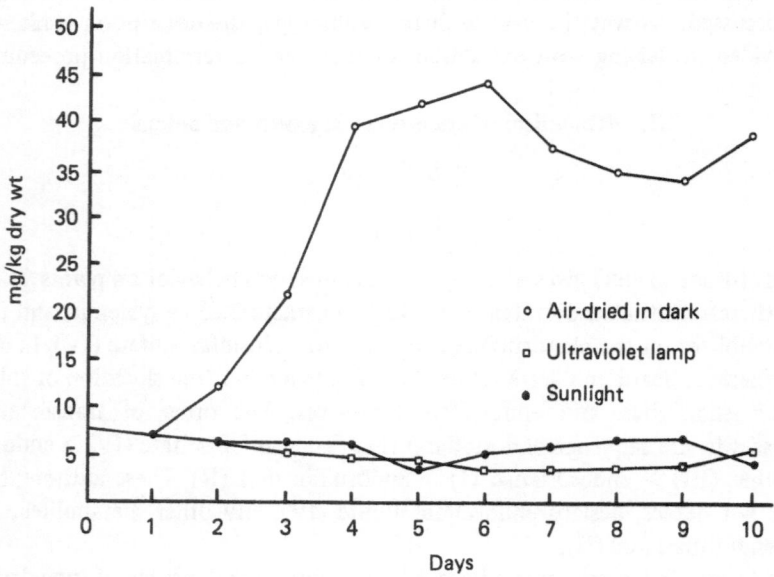

Fig. 3. Changes in endosulfan sulphate on alfalfa with time (after Archer 1973).

minor degree. Strain E53 *Penicillium commune* is at the head with 35% conversion, followed by strain E21, c.f. *Papularia sp.*, with 28%.

In contrast to fungi, bacteria degrade endosulfan (I) preferably to endosulfan diol (II). The strain 295 *Nocardia* sp. stands with 77% at the head, followed by 14 *Corynebacterium* sp. (59%) and J. B. I. *Bacillus polymyxa* (55%).

The strain 260 *Corynebacterium* sp., however, produces only 5.2% endosulfan diol (II) and 56% endosulfan sulfate (IV). It is similar to *Mycobacterium* sp., strain 111, which converts 50% of the given endosulfan (I) to endosulfan sulfate (IV).

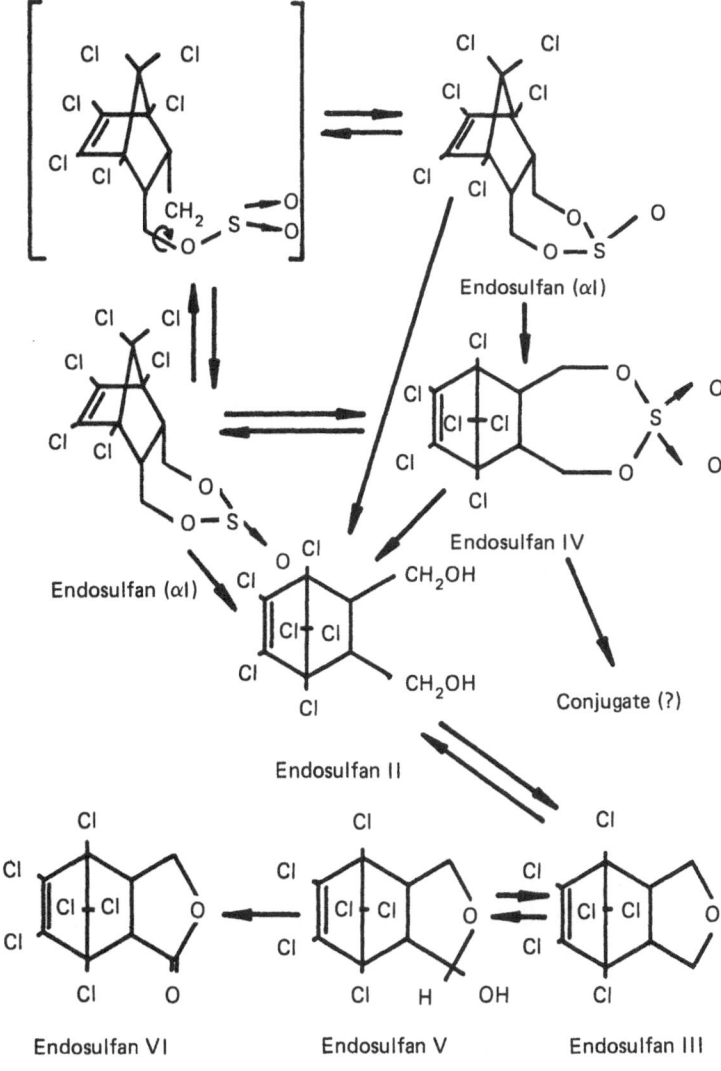

Fig. 4. Metabolism of endosulfan in tobacco leaf (Chopra and Mahfouz 1977).

The streptomycetes, too, are primarily formers of endosulfan diol (II). The microbial degradation of both of the endosulfan isomers (I) by *Pseudomona-daceae* in water has been studied by Ballschmiter *et al.* (1973), who confirmed the results reported previously by other authors (Gorbach and Knauf 1971, Martens 1972), according to which endosulfan (I) is strongly absorbed by microorganisms and is degraded predominatly to endosulfan diol (II). β-endosulfan (βI) is isomerized in a minor proportion to α-endosulfan (αI) and degraded only scarcely to endosulfan ether (III).

El Zorgani and Omer investigated in 1973 the degradation of endosulfan (I) by *Aspergillus,* which they isolated from a Sudanese soil sample. Cultivated in a liquid culture, this fungus degrades a 34% of the given endosulfan (I) to endosulfan diol (II).

A complete degradation scheme of endosulfan (αI, βI) by using a mixed culture of soil organisms, as compared with the degradation under sterile conditions, was presented by Miles and Moy (1979) (cp. Fig. 5).

The half-life periods of the individual metabolites in sterile and in inoculated nutrient media are compiled in Table V.

c) Insects

Rückert and Ballschmiter (1972) continued the studies carried out by Ballschmiter and Tölg (1966) and reported that, contrary to what was believed so far, endosulfan lactone (VI) could not be considered as final product of conversion in insects, but that the oxidative degradation continues. As a further link in the chain of higher oxidized compounds—indirectly over hydrolysis and etherification—a compound could be gas-chromatographically identified, to which the structure of hydroxyendosulfan lactone (VII) is ascribed.

For the oxidative degradation of endosulfan (I), Rückert and Ballschmiter (1972) presented the scheme shown in Figure 6.

In 1972, Uk and Himel reported on the rate of degradation of endosulfan (I) in houseflies (*Musca domestica*) and stated a half-life period of 176 min. The thorax cholinesterase of *Musca domestica* is not inhibited by endosulfan (I).

d) Warm-blooded animals

Gorbach *et al.* (1968) found that a single administration of 0.3 mg/kg ^{14}C-endosulfan (IA) to two lactating milk sheep is almost completely excreted within 22 days. Approximately 50% of the administered ^{14}C activity are excreted with the feces, 41% with the urine, and approximately 1% with the milk. In the total milk of the 22nd day after starting the test, the ^{14}C concentration in terms of endosulfan (I) is as low as 2 μg/kg. At the end of the test period (40 days after administration of the active substance) the ^{14}C concentrations in terms of endosulfan are below 0.03 mg/kg in the meat, fat, and the organs.

The administered endosulfan (I) is not completely metabolized. Intact active substance occurs in the feces, yet not in the urine. Of the metabolites found in

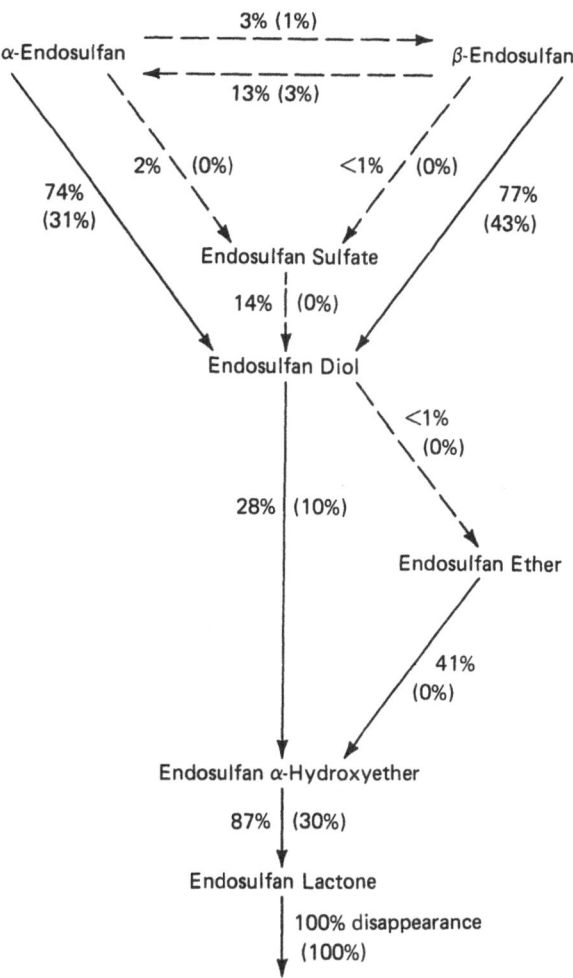

Fig. 5. Conversion of α- and β-endosulfan and metabolites in aqueous nutrient medium inoculated with a mixed culture of soil microorganisms and in sterile medium (bracketed numbers) (Miles and Moy 1979).

the urine, one was found to be endosulfan diol (II), a second proved to be hydroxyendosulfan ether (V). The water-soluble metabolites still present in the urine were not identified.

After acid hydrolysis, a further compound is also extractable, with a behavior like that of endosulfan diol (II).

Schuphan *et al.* (1968) reported on experiments with ^{14}C endosulfan. The tests were intended to check the findings of Deema *et al.* (1966) as to the presence or absence, respectively, of endosulfan diol (II) as hydrolysis product. An amount not given in detail of endosulfan (αI) and endosulfan (βI), respectively,

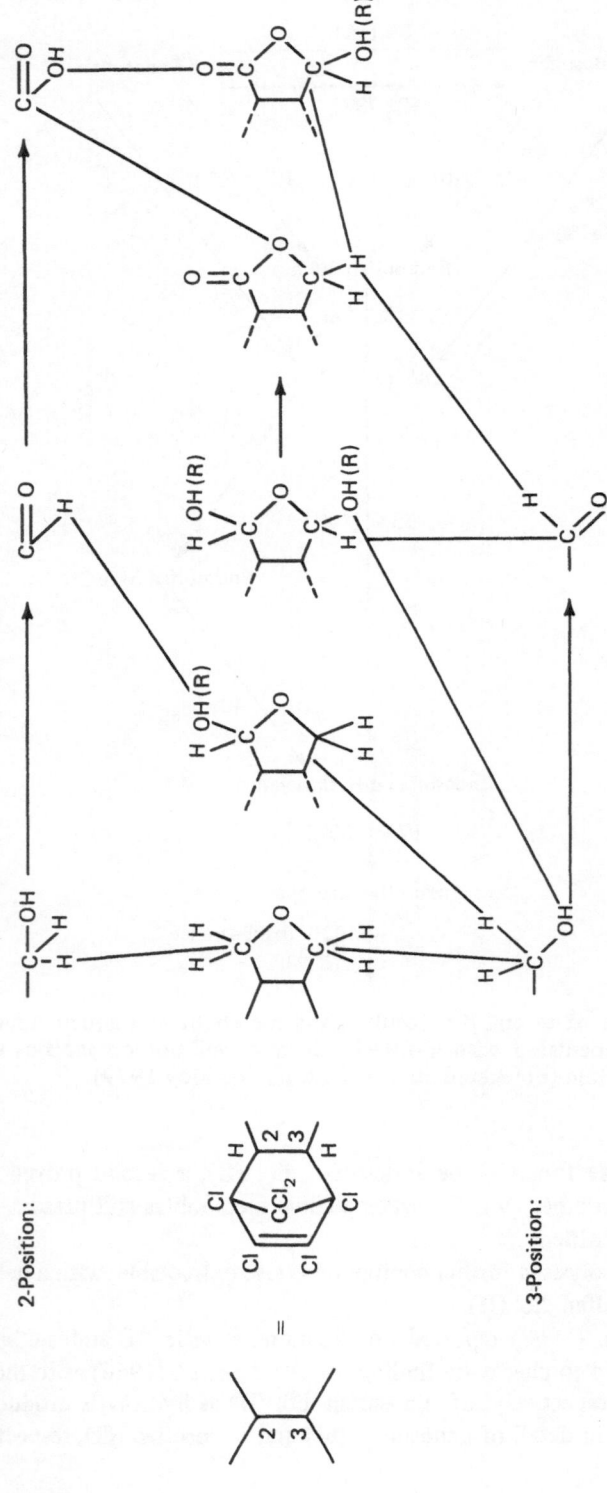

Fig. 6. Scheme of oxidative degradation of hexachlorobicyclo[2.2.1]hept-5-ene derivatives with methylene groups in 2- and 3-position (Rückert and Ballschmiter 1972).

Table V. *Stability of endosulfan metabolites in sterile nutrient (control) medium, and in medium inoculated with mixed culture of soil microorganisms* (Miles and Moy 1979).

	Time for 50% degradation	
	Control	Inoculated
Endosulfan lactone	5.5 hr	5.5 hr
α-endosulfan	12.5 wk	1.1 wk
β-endosulfan	5.7 wk	2.2 wk
Endosulfan ether	> 20 wk	6 wk
Endosulfan α-hydroxyether	> 20 wk	8 wk
Endosulfan sulfate	> 20 wk	11 wk
Endosulfan diol	> 20 wk	14 wk

is fed to mice, and in a second test 4 mg/kg ^{14}C endosulfan (IA) are administered intraperitoneally. Strong excretion of the endosulfan isomers (αI + βI) as well as of endosulfan lactone (VI) in the feces is found. With the urine, primarily endosulfan lactone (VI) and a not yet identified metabolite are excreted. The endosulfan sulfate (IV) and hydroxyendosulfan ether (V) occur in the feces. The said metabolites designated as lipophilic metabolites amount to approximately 20% of the total metabolites present.

Dorough *et al.* (1978) administered endosulfan (αI) and endosulfan (βI) separately orally to rats, on the one part in the form of a single dose, and on the other part on 14 consecutive days. No significant difference as to the degradation behavior of the two isomers was observed. Five days after the single application 75% of the dose was excreted with the feces and 13% with the urine. In the long-term feeding test over 14 days the excretion rates immediately after stopping administration of the active substance amounted to 56% (feces) and 8% (urine). As compared with polar metabolites, the content in apolar metabolites [endosulfan sulfate (IV), hydroxyendosulfan ether (V), endosulfan lactone (VI), and endosulfan ether derivatives] was low both in the excretions and in the tissues. These authors, too, observed the rapid drop of the residue content in the organs after termination of the endosulfan diet (cp. Table VI).

A further pharmacokinetic study was carried out by Gupta and Ehrnebo (1979). Endosulfan (αI + βI) was in this case administered intravenously to rabbits (2 mg/kg). The blood plasma time-content dependence could be correlated for endosulfan (αI) at best for a three-step degradation model. The last-standing regression straight line had an inclination which allowed of estimating the half-life period of the third excretion process of 235 ± 168 hr (standard deviation at N = 6 measurements).

As to endosulfan (βI) the correlation to a two-step degradation suggested itself (half-life period of the second step 5.97 ± 2.41 hr). Gupta and Ehrnebo (1979) conclude from these and further findings give in the study that the two isomers (αI, βI) presented a significant difference in their pharmacokinetic profiles.

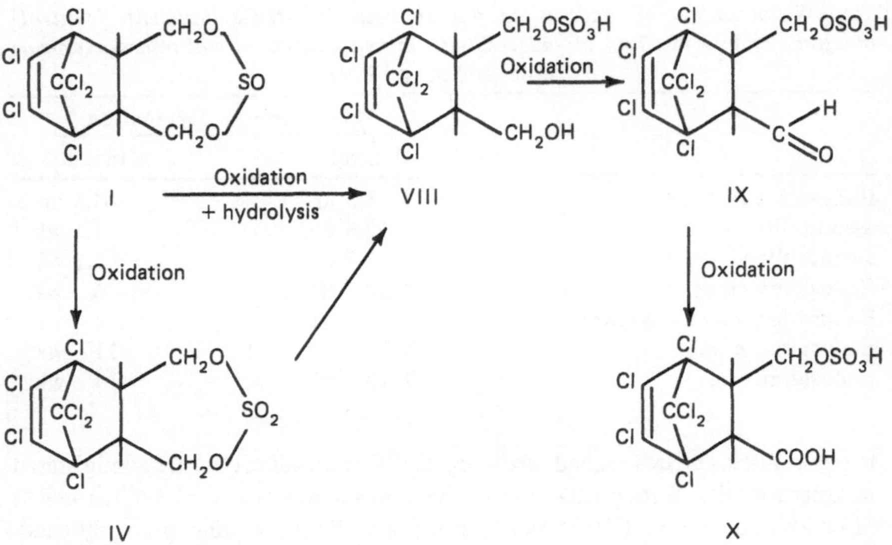

Fig. 7. Degradation of endosulfan (after Elzner 1973).

The experiments with ^{35}S-labeled active substance (ID) showed that the sulfur is more readily cleaved off from endosulfan (βI) than from endosulfan (αI), since in these tests ^{35}S is detected in the urine as [35]SO$_4^{--}$.

Maier-Bode (1968) reported summarily on the degradation of endosulfan (I) in farm animals.

McCaskey and Liska (1967) fed milkers 0.5 to 1 g of technical endosulfan daily during a period of 7 to 11 days and collected the daily milk yield for analysis. Endosulfan (I) itself, the parent compound, was not found in the milk but merely the oxidation product endosulfan sulfate (IV).

Schmidlin-Mészáros and Romann (1971) came to the same result. The milk of cows was examined after they had grazed on pastures accidentally treated with endosulfan (I) (8 to 900 mg/kg of endosulfan in the grass). The milk of said animals contained less than 6 µg/kg of endosulfan (I).

The tests with ^{14}C-labeled endosulfan revealed that fishes are capable of forming water-soluble endosulfan metabolites in the liver. Analyses suggest that endosulfan diol (II) is formed, which is conjugated with glucuronic acid and is passed via the bile to the feces and excreted (Schoettger 1970).

Summary

The degradation of endosulfan in plants, microorganisms (fungi, bacteria, and algae), insects, and warm-blooded animals is reported.

Table VI. *Residues in tissues of female rats fed 5 mg/kg of α- or β-[^{14}C]endo-sulfan in the daily diet.*

	mg/kg of [^{14}C] endosulfan equivalents/isomer in diet[a]									
	Kidney		Liver		Visceral fat		Subcutane-ous fat		Muscle[b]	Brain[b]
Days	α	β	α	β	α	β	α	β	α/β	α/β
On treatment										
1	0.38	0.47	0.26	0.32	0.34	0.24	0.32	0.30	0.02	0.03
2	1.26	1.21	1.02	0.79	0.85	1.02	0.23	0.34	0.02	0.03
7	1.77	1.87	0.96	0.75	0.74	0.53	0.51	0.30	0.02	0.04
10	2.28	2.08	1.11	0.94	0.94	0.55	0.15	0.28	0.03	0.04
14	3.00	3.26	1.08	1.06	0.62	0.50	0.15	0.32	0.05	0.07
Off treatment										
1	2.75	3.34	1.00	0.87	0.45	0.42	0.02	0.08	0.05	0.05
3	1.89	2.21	0.49	0.57	0.15	0.28	0	0	0.02	0.06
7	1.53	1.66	0.28	0.36	0	0	0	0	0	0.04
14	0.94	0.92	0.11	0.19	0	0	0	0	0	0.02

[a] Zero indicates residues were less than 0.02 mg/kg the limit of detectability.
[b] These low residues were representative of both α and β treatments (Dorough *et al.* 1978).

Other studies dealing with the distribution of endosulfan (I) in the cat's brain and the rat's brain, respectively, were conducted by Khanna *et al.* (1979) and Gupta (1978), respectively. In cats, endosulfan, after intravenous administration of endosulfan (I), reached after 15 min a maximum in the lipids of the cerebral cortex and in the cerebellum and thus attained the triple concentration of that found in the brain-stem lipids.

Elzner (1973) more closely investigated the nature of the metabolites occurring in the rat's urine by using ^{35}S (ID) and ^{14}C (IB) labeled endosulfan and assumed two ways of degradation for endosulfan:

1. Oxidation and hydrolysis of endosulfan to hydrophilic sulfur-containing metabolites, acid sulfuric acid esters of endosulfan diol (II) and its oxidation products.

    As hydrolysates of the sulfur-containing metabolites, endosulfan diol (II), hydroxyendosulfan ether (V), and endosulfan lactone (VI) could be identified by gas chromatography and thin-layer chromatography.

    As lipophilic metabolites, endosulfan sulfate (IV) is excreted (cp. Fig. 7).

2. Hydrolysis of endosulfan (I) to endosulfan diol (II), partly oxidation to hydroxyendosulfan ether (V), and endosulfan lactone (VI), and conjugation of these substances to hydrophilic excretion products (cp. Fig. 8).

    Endosulfan lactone (VI) and endosulfan sulfate (IV) are excreted as lipophilic metabolites.

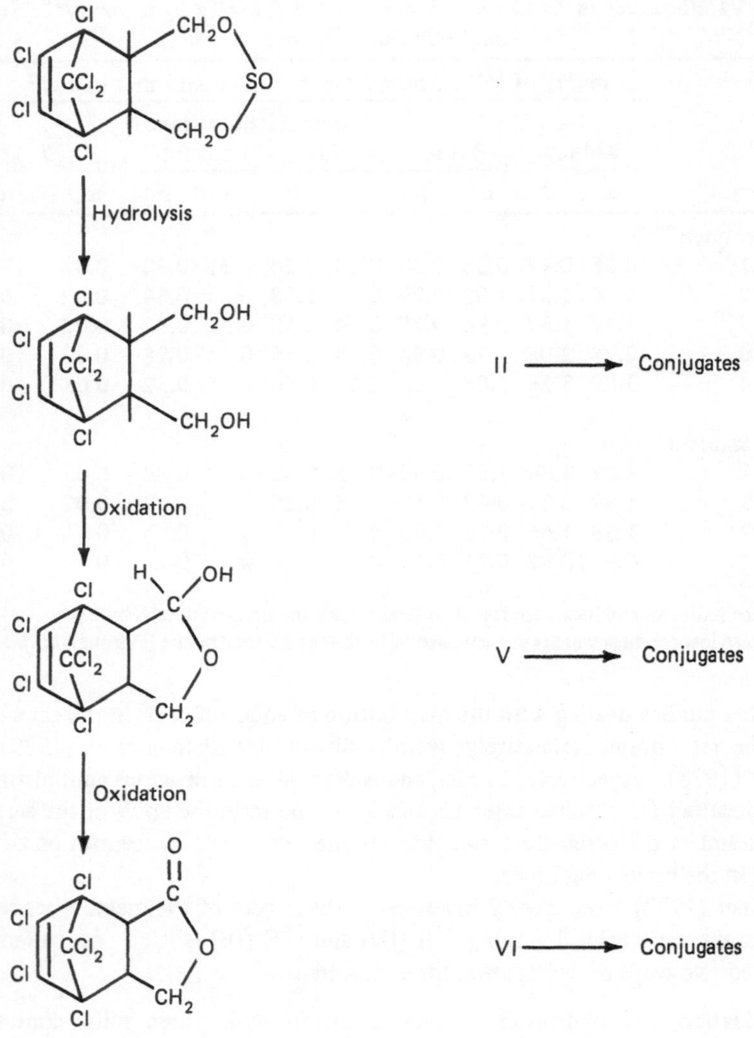

Fig. 8. Degradation of endosulfan (after Elzner 1973).

From endosulfan (I), there form as metabolites preferably endosulfan diol (II), endosulfan sulfate (IV), furthermore endosulfan lactone (VI), and hydroxyendosulfan ether (V). Endosulfan (I) itself is not stored in body fat. Only minor amounts of endosulfan sulfate (IV) are found in the organs.

IV. Toxicity of endosulfan and its metabolites

By R. H. Rimpau

a) Introduction

Since the publication of Maier-Bode (1968), the toxicity of endosulfan, its metabolites, as well as of endosulfan formulations has been repeatedly elaborated in different institutes of several countries on vertebrate species. With endosulfan, especially long-term studies were carried out and the results were again discussed, in line with the recommendations issued by *World Health Organization* (WHO 1968), also in the corresponding bodies of FAO and WHO (WHO 1969 and 1972). On the part of the *WHO Expert Committee on Pesticide Residues*, these reconsiderations did not leave open any toxicological questions concerning endosulfan.

b) Two-year feeding test

As a result of chronic two-year feeding tests in rats, strain Wistar, technical grade endosulfan was shown to have a no-effect level of 30 mg/kg feed (*Hazleton Laboratories, Inc.* 1959). On administration of 3 mg/kg of technical endosulfan per kg diet to dogs over the same period, said dose proved to constitute the no-effect level (*Industrial Bio-Test Laboratories, Inc.* 1967).

c) Three-generation reproduction study

In a three-generation reproduction test in rats, strain Sprague-Dawley, no substance-conditioned alterations attributable to the feeding of 2 or 50 mg/kg diet, respectively, of endosulfan technical in the daily diet could be found in all three generations, neither in the dams (F_2) nor in their offspring (F_{3a} and F_{3b}) (*Industrial Bio-Test Laboratories, Inc.* 1965).

d) Teratogenicity test

Another study by *Industrial Bio-Test Laboratories, Inc.* (1972) deals with the examination of technical endosulfan for the existence of any teratogenic effect. Endosulfan technical was administered to rats, strain Charles River, in doses of 0.5 and 1.5 mg/kg body wt/day from the 6th to 15th day of pregnancy. All investigated parameters: body weight of mothers, mortality of mothers, number of implantation sites, number of resorption sites, viable fetus, obvious fetal deformities, fetal skeleton development, as well as internal development, did not reveal any significant differences between test animals and controls.

e) Cancerogenicity tests

In a cancerogenicity test, mice were fed for lifetime 3 mg or 6 mg of endosulfan technical per kg diet, respectively, which did not lead to any significant rise of the tumor rate (Innes et al. 1969).

The cancerogenicity tests published by National Cancer Institute, National Institutes of Health, U.S. Department of Health, Education, and Welfare (1978), carried out with endosulfan technical in female rats and female mice, did not suggest any cancerogenic properties.

f) Mutagenicity test

Endosulfan was administered orally to rats at 0, 11.00, 22.00, 36.60, and 55.00 mg/kg daily for 5 days. The highest doses were associated with clinical signs of insecticide poisoning and death. Cytogenetic analyses of bone marrow cells and spermatogenetic cells did not reveal any significant effect of the insecticide on chromosomes. The ratio of mitotic index and frequency of chromatid break in the two cell types had no correlation with the doses tested and was not very different from those of the control group (Dikshith and Datta 1978).

g) Acute oral toxicity of the isomers and metabolites

The following Table VII is a compilation of the test results on acute oral toxicity of α- and β-endosulfan, as well as of their metabolites, in rats.

Summary

In accordance with the report issued by WHO (1972) it can be confirmed in summary that for endosulfan, no open questions or unsolved test problems have remained with regard to toxicity, fertility, teratogenicity, cancerogenicity, as well as to mutagenicity.

V. Environmental toxicology of endosulfan and its metabolites

By W. Knauf

a) Introduction

A series of test results on possible side-effects of endosulfan on the biosphere reveals that endosulfan, on account of its biological properties and its chemical behavior, significantly differs from the class of chemicals designated as "chlorinated hydrocarbons".

The following is a compilation of test results with endosulfan in a series of organisms including—as far as is known—observations in field tests, e.g., from the application of endosulfan as Thiodan®.

b) Aquatic organisms

1. Heterotrophic microorganisms (bacteria and soil fungi).—Various bacteria strains show a high tolerance to endosulfan. In TCC tests as well as in similar tests with fermentation tubes, the injury threshold was in a range of 250 to 500 mg/kg (Weidenmueller et al. 1971), viz., invariably far above the point of maximum solubility of endosulfan in water.

It could further be demonstrated (Martens 1972) that at a nominal concentration of 1,000 mg/kg of endosulfan, no selection of the single strains occurred in particular culture media after inoculation with a mixture of different soil fungi; no death of specific strains could be observed. These investigations covered 28 different soil fungi. Tests carried out by the same author with 50 different species of bacteria and fungi (*Streptomyces*) gave similar results. Not even a decline in biomass production (determined by wt increase of microorganisms at a concentration of 100 mg/kg endosulfan (I)) was observed. This signifies that soil microorganisms are not adversely affected even by an overdose of endosulfan.

2. Autotrophic microorganisms.—

α. *Unicellular green algae.*—Investigations on the influence of endosulfan (I) on green algae with *Chlorella vulgaris* as example were also extended so as to cover the metabolites endosulfan alcohol (II), endosulfan ether (III), endosulfan sulfate (IV), hydroxyendosulfan ether (V), and endosulfan lactone (VI) (Knauf and Schulze 1973 a). The tests were conducted under long-term conditions (120 hr of treatment comprising approximately 20 generations) in culture thermostats under daily renewal of the culture solution. Under these conditions, the photosynthetic efficiency of the algae was not impaired before a concentration of 10 mg/kg (measured as O_2-production in a Warburg apparatus). In the observation of the rhythm of cell division, a decrease of the rate of division can only be evidenced at a concentration of more than 2 mg/kg. This is 2,000-fold the concentration found in nontreated surface waters during intensive application in rice cultivation areas of Indonesia (Gorbach et al. 1971 a). The said effect proved to be fully reversible after termination of the contamination of the culture solution.

Test-tube investigations with a very high endosulfan concentration were carried out during 14 days with *Selenastrum*, another unicellular alga. Injury to the algae only became apparent at a concentration > 80 mg/kg (Knauf and Schulze 1973 b).

β. *Filamentous green and bluegreen algae.*—Tests with filamentous algae have been carried out with blue as well as green algae. It was found that endosulfan very rapidly penetrates the cells of *Cladophora*; within one hr and a half of the initial endosulfan content of the culture solution could be recovered in the test algae (30 to 200 μg/kg) (Bauer 1972).

The same could be shown with the unicellular green alga *Chlorella vulgaris* at still higher concentrations (50 mg/kg) (Oeser et al. 1971).

Table VII. *Acute oral toxicity (LD₅₀-values) of α- and β-endosulfan and of endosulfan metabolites in rats.* [a]

Substance	Structural formula	LD_{50} in mg/kg body wt. [b]	Date of report
α-Endosulfan (αI)		76	8/25/1975
β-Endosulfan (βI)		240	8/26/1975
Endosulfan sulfate (IV)		76	5/29/1964
Endosulfan ether (III)		>15,000	1/31/1964

Endosulfan hydroxyether (V)	1,750	9/14/1966
Endosulfan lactone (VI)	165-290 105-115[c]	8/25/1975 5/21/1971
Endosulfan diol = Endosulfan alcohol (II)	> 15,000	1/31/1964

[a] Investigating laboratory: Pharma Research Toxicology, Hoechst AG.
[b] Vehicle = starch mucilage.
[c] Vehicle = sesame oil.

After the uptake of endosulfan in the cell body, a strong holding-back effect can be observed. The tests demonstrated that the green algae are capable of converting endosulfan after uptake first to endosulfan alcohol, which is again given off. By this process the toxicity to aquatic organisms is also drastically reduced.

In further experiments in which the blue alga *Phormidium* spp. was exposed for a period of 14 days in a test-tube to the substances endosulfan (I), α-endosulfan (αI), β-endosulfan (βI), endosulfan diol (II), endosulfan ether (III), endosulfan sulfate (IV), hydroxyendosulfan ether (V), or endosulfan lactone (VI) in a concentration of 1.25 mg/kg, no alteration of the color of the algal filament could be detected on visual estimation of the respective cultures (Knauf and Schulze 1973 b). At 20 mg/kg a change of shade of the algae towards gray could be ascertained with substances possessing sulfur in the molecule [endosulfan (I), α-endosulfan (αI), β-endosulfan (βI), endosulfan sulfate (IV)]. Metabolites devoid of sulfur in the molecule [endosulfan alcohol (II), endosulfan ether (III), hydroxyendosulfan ether (V), and endosulfan lactone (VI)] did not yet cause any color change of the alga culture at 20 mg/kg.

The above-cited test results add to the knowledge gained with higher plants concerning the extremely low phytotoxicity of endosulfan.

3. **Aquatic animal organisms.**—

α. *Insects, crayfish, molluscs (fish feed animals)*

αα. *Acute toxicity = laboratory investigations.*—Toxicity investigations with a multitude of aquatic organisms other than fish have been carried out by several authors (*e.g.*, Knauf and Schulze 1973 a, Bauer 1961, Schoettger 1970, Herbst 1971).

Results hitherto obtained by said authors are compiled in Table VIII. They indicate that the LC_{50} values are mostly $\geqslant$ 0.1 mg/kg, frequently even $\geqslant$ 1 mg/kg. It may be pointed out in this connection that immediately after the application of endosulfan in paddy fields of Java an endosulfan concentration of 0.2 to 0.5 mg/kg was found in the water of those fields, which, however, dropped to a value of < 0.001 mg/kg within three days (Gorbach *et al.* 1971 b).

ββ. *Short-term massive doses of endosulfan in flowing waters and their influence on fish feed animals.*—In a model study van Dyk (1977) investigated the influence of short-term high endosulfan concentrations in a slowly flowing stream in South Africa. The rivulet held a rich aquatic fauna and was not polluted by industrial waste waters. Endosulfan was applied in a single dose of 90 mg active ingredient (a.i.) into the rivulet (flow vol. 14,400 L/hr $\hat{=}$ 4 L/sec; flow rate 0.58 to 0.75 m/sec). Samples of the fauna and water were taken immediately after application as well as 3 and 6 hr later at 6 different locations below the place of application. The author indicated that an effect caused by endosulfan on the number of animals found is likely but not quite sure. Immediately after application a reduction of the number of organisms/m^2 could be ascertained at the sampling places quite near the application site. These values, however, one to two days after treatment rise rather rapidly. It is interesting to note that after

Table VIII. *The effect of endosulfan on fish food organisms (static test).*

Species	LC-values	mg/kg
Insecta		
Aedes aegypti	LC_{50} (48 hr)	0.08
Chironomus plumosus	LC_{50} (48 hr)	0.025
Cloeon dipterum	LC_{50} (68 hr)	1.0
Nemourella picteti	LC_{50} (72 hr)	1.0
Ischnura sp.	LC_{50} (120 hr)	0.06-0.07
Crustacea		
Daphnia magna	LC_{50} (48 hr)	0.24
Daphnia pulex	LC_{50} (24 hr)	0.3
Artemia salina	LC_{50} (48 hr)	9.5
Cambarus affinis	LC_{50} (24 hr)	0.5
Cyclops strenuus	LC_{100} (24 hr)	1.0
Asellus aquaticus	LC_{50} (24 hr)	0.1
Mollusca		
Planorbis corneus	LC_{50} (24 hr)	1.2
Limnaea stagnalis	LC_{50} (24 hr)	1.6
Physa sp.	LC_{50} (24 hr)	0.21
Ancylus sp.	LC_{50} (11 d)	0.01

treatment, with decreasing distance from the contamination point, the number of organisms/m^2 increases (considered in counting: *Copepoda, Ostracoda, Ephemeroptera, Chironomidae, Simuliidae, Corixidae, Notonectidae, Hydrophylidae,* and *Hydreanidae*).

γγ. *Investigations of the influence of endosulfan on fish feed animals in areas of Thiodan application.* –

In South Africa, down-stream of the Loskop dam situated in an area of intensive endosulfan application in cotton, van Dyk (1977) carried out population density determinations on a series of aquatic insects. The investigations covered the following insect orders (suborders):

Crustacea	*Cladocera*
	Copepoda
	Ostracoda
Hemiptera	*Notonectidae*
	Belostomatidae
	Gerridae
	Corixidae
Diptera	*Simuliidae*
	Chironomidae
Ephemeroptera	*Ephemeridae*

The concentration of endosulfan $\alpha + \beta$ in the waters of the rivers of the Loskop dam area reached in January/February (1974) a level of 1.1 to 4.1 μg/kg. In the rivers of the Loskop dam irrigation area, the established population densities at that time were found to be rather low ($<$ 8,000 organisms/m^2), a fact accounted for by the author by a seasonal change in the habitats of the organisms. According to the author's opinion, endosulfan is not responsible for the fluctuation of the population density; nor could any correlation be ascertained between the recovered endosulfan residues and the rise or decline of the population density of the aquatic fauna in tissue.

δδ. *Long-term effect on Daphnia magna (life-cycle test) and Mytilus edulis.*— In order to study the influence of plant pesticides upon the reproduction of animal organisms, the so-called life-cycle test with the water flea *Daphnia magna* is increasingly used.

To this effect, 24-hr-old organisms are kept in glass beakers as single specimens in a test medium, and they are periodically conveyed into freshly contaminated media. Hence the organisms are exposed to roughly equal concentrations of the substance to be examined (semi-static test). With endosulfan, such a test was carried out for 35 days with *Daphnia magna* (*Hoechst*, internal report 1980).

Containers with 100 ml of test water were used; the *Daphnia* were transferred separately every 7 days by hand into a new test medium. Daily determination was conducted concerning: mortality of parent organisms, number of produced offspring/animal, and mortality of newborn young. The endosulfan concentrations were 0.01 and 0.001 mg/kg (applied as Thiodan 35 fluid).

During the whole period of observation (35 days), neither in the mortality of the parent organisms nor in the number of produced young/day was any difference to the control series to be seen. Merely in the mortality of the young animals (summed up over the whole period of observation) was a moderate increase ascertained for the mortality at 0.01 mg/kg, and a slight increase at 0.007 mg/kg.

Investigations with *Mytilus edulis* (mussel) in salt-water in a flow system over 112 days at increasing concentrations (up to 1 mg/kg) revealed that at this concentration the biological activity of the mussel remains unchanged during the entire observation period (Roberts 1972).

β. *Fish.*—

αα. *Acute toxicity.*—Figures concerning the endosulfan toxicity and that of its metabolites towards fish are presented in Table IX. Generally it should be stated that the LC_{50} values for endosulfan lie within the concentration range of 0.01 to 0.001 mg/kg.

It could be established in investigations with the metabolites that sulfur-free metabolites (II, III, V, VI, and VII) possess a significantly lower toxicity against fish (Knauf and Schulze 1973 a). The LC_{50} values of said substances are in the range 1 to 10 mg/kg.

The effect (*i.e.*, the rate of action) of endosulfan in aquatic organisms is temperature-dependent. Macek *et al.* (1969) reported a decline of the endosulfan

Table IX. *The effect of endosulfan on fish.*

Species	LC-values	mg/kg
Salmo gairdneri	LC_{50} (58 hr)	0.0021-0.0011
Esox lucius	LC_{100} (48 hr)	0.015
Cyprinus carpio	LC_{50} (48 hr)	0.011
Carassius auratus	LC_{50} (48 hr)	0.01
Lebistes reticulatus	LC_{50} (48 hr)	0.0015
Xiphophorus helleri	LC_{100} (24 hr)	0.005-0.01
Puntius javanicus	LC_{100} (48 hr)	0.001
Catastomus commersoni	LC_{50} (48 hr)	0.0064-0.0043
Idus idus	LC_{50} (48 hr)	0.0046
Leucaspius delineatus	LC_{50} (16 hr)	0.02
	LC_{50} (96 hr)	0.005
Gobio gobio	LC_{50} (24 hr)	0.005-0.007
Rutilus rutilus	LC_{50} (48 hr)	0.0014
Rasbora heteromorpha	LC_{50} (48 hr)	0.00009

action on *Salmo gairdneri* (rainbow trout) at low temperatures. The influence of temperature decreases with the prolongation of the time of exposure, which suggests an influence of the temperature on the rate of action (Table X).

Higher salt contents of water do not change the toxicity of endosulfan against aquatic organisms (*Catastomus commersoni*). Studies with *Idus idus* (golden orfe) have confirmed this fact (Knauf and Schulze 1973 a, Schoettger 1970).

ββ. Sublethal effects.—van Dyk (1977) investigated in a special test arrangement the swimming capacity of *Sarotherodon mossambicus* under the influence of endosulfan. The test apparatus consisted of a container ($1.50 \, cm \times 10 \times 20 \, cm$) into which endosulfan-containing water (0.1 to 0.5 µg/kg) was pumped (45 hr) from a storage vessel. The water temperature was 24°C. By means of photocells the movements of the fish in a channel could be automatically recorded. In comparison with the controls, the mobility of the test animals was increased up to seven-fold in all tested concentrations (0.1, 0.2, 0.3, 0.4, and 0.5 µg/kg).

γγ. Growth and fertility investigations.—The same author studied the catch weight of *Sarotherodon mossambicus* in endosulfan-contaminated waters of the Loskop dam area in South Africa. With 84 caught adult specimens, the mean

Table X. *The LC_{50}-values of endosulfan at different temperatures in Salmo gairdneri (after Macek et al. 1969).*

°C	LD_{50} (µg/kg)	
	24 hr	96 hr
1.6	13 (11-15)	2.6 (2.3-2.9)
7.2	6.1 (5.6-6.6)	1.7 (1.5-1.9)
12.7	3.2 (2.9-3.5)	1.5 (1.3-1.7)

weight was 182 g/fish, which did not significantly differ from the average fish weight from untreated waters. According to the author's conclusion, the growth of the fish, therefore, is not influenced by sublethal concentrations of endosulfan. The production of young was 73 per female among the treated fish, which amounts to a low production. During the test period, however, predatory fish of the variety *Micropterus salmoides* appeared in the first year, and in the second year another kind of predator reptile (*Iguanas*), which, without doubt, affected the figures concerning the reproduction rate.

δδ. *Pathological alterations in fish due to endosulfan intoxication.*—Walsh (1975) investigated *Onchorhynchus kisutch* and *Salvelinus namaycush* for pathological alterations after exposure of the test fish over a period of 7 days to endosulfan (the author used as test concentration such that would have sufficed under the given conditions to kill the fish within 30 days). No histopathological or pathological alterations, respectively, of diagnostic value were found (which was in accordance with parallel tests of other plant pesticides).

εε. *Influence on enzyme activity in fish.*—*In vitro* investigations for inhibition of the ATPase activity in the cerebral tissue of *Lepomis macrochirus* (Blue gill sunfish) were carried out by Yap *et al.* (1975). Under the chosen conditions (concentration 20.8 μM at 37°C), a medium to strong inhibition of the sensitive and insensitive oligomycin Mg^{2+} ATPase was found. The $Na^+ - K^+$ ATPase, as compared with the control, was less strongly inhibited. In the inhibition of mitochondrial Mg^{2+} ATPase, the next to lowest value in 6 tested compounds was found with 14.9 μM (ID_{50}).

Dalela *et al.* (1978) investigated the influence of sublethal long-term exposure to endosulfan in *Channa gachna*, a sweet water teleost from India, *in vivo*. The authors exposed animals of the said kind to concentrations of 5.34, 3.55, 2.13, and 1.74 μg (respectively 1/2, 1/3, 1/5, and 1/6 of the LC_{50} during 96 hr with this species) for 30 and 60 days under flow conditions. These are extremely high concentrations (as compared with the numerical difference to the LC_{50}, 96 hr), so that a strong intoxication was to be expected.

In higher dose levels the tests showed inhibitions of tissue enzymes in graded intensities. Brain, gills, liver, and kidneys were examined. In the lowest tested concentration (of 1/6 of the LC_{50}), the values for the Na^+K^+ ATPase differed but slightly, if any, from the controls. A significant inhibition—especially in the highest concentration tested—could be observed, however, with the oligomycin-nonsensitive Mg^{2+} ATPase in the cerebral, gill, and liver tissue after 60 days. In the kidneys, however, the strongest inhibition was observed with the oligomycin-sensitive (mitochondrial) Mg^{2+} ATPase. The authors conclude from this that endosulfan has an influence on the various energy-dependent processes in the fish tissue.

c) Terrestrial organisms

There are many observations on the effect of endosulfan on a variety of terrestrial organisms.

1. Plants.—Endosulfan is known to be extremely plant-tolerant. In a very great number of laboratory and field tests, even in highly sensitive crops, no serious injury has occurred. The information presented by Maier-Bode (1968) is referred to.

2. Animals.—

α. *Invertebrates.*—It is not surprising that an insecticide still effective in such low concentrations, as is the case with endosulfan, possesses a relatively broad spectrum of activity within the insect group. In intensive crops (such as cotton), the said property is expressly desirable. Despite the broad spectrum of activity, endosulfan spares quite a number of so-called beneficial organisms. Emmel (1958) reported in connection with the control of *Eriosoma lanigerum* in fruit culture on observations in wooly aphid parasites *Aphelinus mali*, a well-known beneficial insect a parasite of *Eriosoma lanigerum*. Apple twigs cut in the field bearing parasitized wooly aphids were treated in the laboratory with endosulfan WP or endosulfan EC, respectively; parathion EC served as comparative agent. The wasps hatching from the wooly aphids within a 4-wk period of observation were counted for each test. As compared with untreated controls it became distinctly apparent that in contrast to parathion no injury occurred in *Aphelinus mali* by employing the conventional application rates (Table XI).

In a test in citrus fruit with *Aphytis africanus*, one of the main parasites of the red scale *Aonidiella aurantii*, it could be demonstrated that in the plots treated with endosulfan the parasite potency of *Aphytis africanus* was approximately equal with that in the untreated controls (Pierza and Fisher 1965).

Swirski *et al.* (1968) reported on the action of endosulfan on the predator *Amblyseius swirskii* (*Phytoseidae*). They ascertained that endosulfan causes an initial mortality in *A. swirskii*, but that the population after treatment recovers in such a rapid way that the number of predators/fruit reaches after 135 days two- to three-times that of the controls (Table XII).

Sharma *et al.* (1971) reported on the effect of endosulfan on *Chrysopa* sp. and *Brumus* sp. (predators in cotton crops). According to these authors, endosulfan scored the most favorable result in a comparative test with methyldemeton, formothion, dimethoate, and thiometon. Up to 15 and 7, respectively, days

Table XI. *Influence of Thiodan*® *on Aphelinus mali (wooly aphid parasite)* (after Emmel 1958).

Formula product	% a.i.[a]	No. of hatched *Aphelinus mali*
Thiodan EC	0.035	79
Thiodan EC	0.0175	107
Thiodan WP	0.035	128
Thiodan WP	0.0175	62
Parathion EC	0.0175	0
Control	–	113

[a] a.i. = active ingredient.

Table XII. *Average no. of predators/fruit* (after Swirski *et al.* 1968).

Treatment	Days after treatment				
	0	10	49	88	135
Thiodan 35 EC (0.4 %)	0.92	0.06	1.2	0.84	0.45
Maneb 80 WP (0.12 %)	1.08	0.01	0.03	0.27	0.16
Mancozeb 80 WP (0.12 %)	1.13	0.08	0.08	0.17	0.08
Control	1.54	1.12	0.63	0.23	0.14

following treatment, the mortality of either kind of predators in the endosulfan-treated plots was significantly lower than in the plots treated with the other pesticides. Only thereafter does the mortality rise also with endosulfan, yet even after termination of the tests, live predators were encountered in the ensosulfan-treated plots. The decrease may be primarily accounted for by the natural mortality of the insects (Table XIII).

Moffitt *et al.* (1972) investigated the results of an endosulfan treatment on *Hippodamia convergens* in comparison with other conventional insecticides. *H. convergens* is considered to be an important beneficial insect by said authors, *viz.*, in fruit culture of the western U.S.A. (State of Washington). It could be established in laboratory and field tests that in contrast to comparative agents azinphosmethyl, parathion, carbaryl, diazinon, and DDT, the said beetles were only very slightly affected. The results are shown in Figure 9.

The staphylinidae *Tachyporus hypnorum* and *Philonthus fuscipennis* are, according to Eghtedar (1969), among the most frequent predator beetles in northwestern German rape culture. In tests with selectively acting insecticides in rape the author found that endosulfan had a relatively weak effect against either kind of insects (Tables XIV and XV).

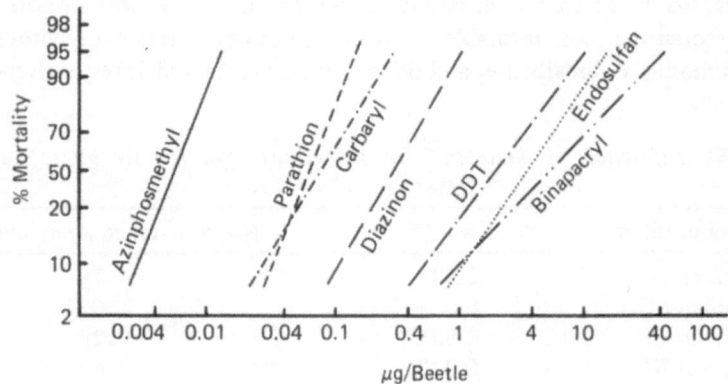

Fig. 9. Dose-mortality graphs of endosulfan and comparative agents in *Hippodamia convergens* (direct application) (Moffitt *et al.* 1972).

Table XIII. *Mortality of Chrypsopa sp. and Brumus sp. at various times after treatment (after Sharma et al. 1971).*

Insecticides	ml/ha	Chrypsopa sp. (% mortality)					
		24 hr	48 hr	72 hr	7 d	15 d	30 d
Methyldemeton 25% EC	1,125	62.47	85.97	92.00	100.00	100.00	100.00
Formothion 25% EC	1,125	44.41	85.97	96.00	100.00	100.00	100.00
Dimethoate 30% EC	875	73.99	81.97	85.97	94.50	100.00	100.00
Dimethoate 30% EC	875	74.50	84.50	94.50	100.00	100.00	100.00
Endosulfan 35% EC	1,500	15.07	23.46	27.96	40.91	65.47	80.00
Endosulfan 35% EC	1,500	20.54	8.99	11.54	26.04	37.99	84.50
Thiometon 25% EC	1,000	54.91	60.41	82.41	86.94	91.47	100.00
Thiometon 25% EC	1,000	44.98	55.88	78.41	37.47	100.00	100.00
		Brumus sp. (% mortality)					
Methyldemeton 25% EC	1,125	64.94	80.47	89.97	100.00	100.00	100.00
Formothion 25% EC	1,125	48.44	81.97	90.50	100.00	100.00	100.00
Dimethoate 30% EC	875	56.45	81.04	86.31	89.97	100.00	100.00
Dimethoate 30% EC	875	46.54	82.31	96.00	100.00	100.00	100.00
Endosulfan 35% EC	1,500	8.92	16.42	37.57	56.41	81.44	92.05
Endosulfan 35% EC	1,500	7.15	19.36	33.83	40.26	71.01	82.31
Thiometon 25% EC	1,000	34.39	45.78	66.78	76.47	86.50	100.00
Thiometon 25% EC	1,000	40.09	50.95	69.40	78.31	87.81	100.00

Table XIV. *Susceptibility of imagos of Tachyporus hypnorum to some insecticides (no. of test animals: 40)* (after Eghtedar 1969).

No.	Agent	Concentration (%)	Mortality in % after 24 hr	48 hr	72 hr
1	Toxaphene	0.2	20	30	72
2	Endosulfan	0.2	0	0	50
3	Activated methoxychlor	0.3	100	100	100
4	Lindane	0.05	100	100	100
5	Parathion	0.035	100	100	100
6	Diazinon	0.1	100	100	100
7	Dimethoate	0.1	100	100	100
8	Bromophos	0.2	100	100	100
9	Carbaryl	0.15	15	15	15
10	Derris + pyrethrum	0.2	0	0	62.5
11	Control	0	0	0	0

In Germany, endosulfan has been recognized as a plant pesticide without hazard to honeybees. During rape blossom, this preparation may be used without risk to flower-visiting bees. Beran (1970), in his compilation of plant pesticides and their toxicity and harmfulness to bees, designated endosulfan as ". . . remarkably tolerant to bees . . ."

Laboratory investigations showed in topical application of the technical-grade active substance an LD_{50} of 16.14 µg/bee (Atkins *et al.* 1973). The author listed the said substance in the group III ("Relatively nontoxic to Honey Bees").

During practical application of endosulfan in crops in bloom, no raised mortality nor drop of efficiency of beehive bees has ever been observed.

Table XV. *Susceptibility of imagos of Philonthus fuscipennis to some insecticides (no. of test animals: 40)* (after Eghtedar 1969).

No.	Agent	Concentration (%)	Mortality in % after 24 hr	48 hr
1	Toxaphene	0.2	10	10
2	Endosulfan	0.2	5	5
3	Activated methoxychlor	0.3	5	5
4	Lindane	0.05	100	100
5	Parathion	0.035	100	100
6	Diazinon	0.1	100	100
7	Bromophos	0.2	100	100
8	Dimethoate	0.1	100	100
9	Carbaryl	0.15	0	0
10	Derris + pyrethrum	0.2	11.5	11.5
11	Control	0	0	0

Huettenbach (1969) and Singh (1969) have given numerous examples of the properties of endosulfan to spare beneficial insects. There were mentioned representatives of the orders *Hymenoptera, Neuroptera, Diptera,* and *Coleoptera,* that are either not affected by endosulfan or significantly less so than by treatment with comparative insecticides.

β. *Vertebrates.*—A series of toxicity data for evaluating the action of endosulfan on terrestrial vertebrates (mammals, birds) has already been discussed in Maier-Bode (1968), Chapter II (Action). Some additional figures on the toxicity to wildlife animals follow here.

αα. *Birds.*—In birds the action of the preparation is highly dependent upon the carrier substance used. For the Japanese quail (*Coturnix coturnix iaponicus*), the values for the acute oral toxicity (LD_{50}) are 26 and 28 mg/kg (starch slime) or 159 to 201 mg/kg (sesame oil), respectively, of body wt (*Hoechst AG* internal report 1972). For the duck (*Anas platyrhynchos*) the oral acute LD_{50} is 243 mg/kg for male and 205 mg/kg for female animals. For the bobwhite quail (*Colinus virginianus*) the LD_{50} acute oral is 50 or 56 mg/kg for male and female animals, respectively (*Niagara* 1972).

After an extensive application of endosulfan (as Thiodan 35 fluid) in the forest (Schifferli 1967), with an application rate of 0.35 kg active substance/ha, no negative effect on nestlings of songbirds was observed. In the said test, 34 varieties of birds had been counted in the treated area; a significant alteration in the number of breeding pairs could not be determined after the treatment.

In Canada, no intoxication of geese (*Anser anser*) was observed, that could feed on herbage of strawberry plots for 17 days. Endosulfan had been applied twice in intervals as 50 WP with 2 lb/100 gal in the usual application rate (Dustan 1965).

Some of these results are collated in Table XVI.

In contrast thereto, the feeding of endosulfan-treated caterpillars by *Parus major* and *Turdus merula* to nestlings resulted in intoxications (Pryzgodda 1961).

Douthwaite (1980) investigated in a carefully conducted monitoring program during an extended application of endosulfan against tsetse flies in Botswana, its

Table XVI. *The toxicity of endosulfan to some wildfowl species* (after Gupta and Gupta 1979).

Species	Route of administration	Vehicle	LD_{50} (mg/kg)	Reference
Mallard duck	Oral	Peanut oil	200-275	Martin (1968)
Ring-necked pheasants	Oral	Peanut oil	620-1000	
Ring-necked pheasants	Oral	?	620-850	Dewitt *et al.* (1963)
Bobwhite	Oral	?	270	

influence and effect on the kingfisher. This bird lives directly in the area of endosulfan application and feeds on fish. The author could ascertain that birds feeding on fish of the region did not contain any measurable endosulfan residues. The success rate of dip-catching of the kingfisher seemed higher after every application; this effect, however, could not be evidenced after each application. The author did not observe any other change of behavior of the birds under test.

ββ. *Mammals.*—Investigations carried out in the U.S.A. on cows (*Hazleton Laboratories, Inc.*, Falls Church, VA, U.S.A., 1959) have shown that even 30 mg/kg of endosulfan in the fodder (fresh lucerne) during a period of 30 days were tolerated without reaction. It must be remembered that the tolerance for lucerne is 0.3 mg/kg of endosulfan in the U.S.A. (*viz.*, at 1/100 of the concentration used in the test).

Summary

It can be shown that bacteria, fungi, monocellular and filamentous green and bluegreen algae tolerate endosulfan in high concentrations ($\geqslant$ 10 mg/kg). Green algae are able to metabolize endosulfan to endosulfan alcohol.

Figures are given of the toxicity of endosulfan to aquatic invertebrates. Under practical conditions, the impact of endosulfan on these organisms is little, if any.

The specific sensitivity of fish to endosulfan is shown either in acute toxicity tests, in behavioral studies, or in measurements of enzyme activity of fish tissue. Substance-specific histopathological effects could not be established. As for the metabolites, their toxicity against fish is drastically reduced after removal of the sulfur from the molecule (factor of 100 to 10,000).

There are only minor signs of adverse effects on birds under practical conditions.

Endosulfan is safe to bees; numerous other beneficial insects also remain unaffected; examples are given.

VI. The environmental behavior of endosulfan and residue values

By S. Gorbach

a) Factors influencing the residue behavior of endosulfan

1. **Volatility.**—The endosulfan isomers and endosulfan sulfate possess a measurable vapor pressure (*cp.*, Section I). The volatilities are in accordance with the vapor pressures, if the substances are deposited as thin films on inert substrates and are exposed at $22°C \pm 1°C$ to a constant air stream (2 L/min).

Figure 10 shows the substance losses in dependence on time (Beard and Ware 1969). The same authors have also investigated the volatility behavior of endosulfan (αI) and (βI), endosulfandiol (II), endosulfan sulfate (IV), and endosulfan ether (III) on bean leaves (*Phaseolus vulgaris*) and sugarbeet leaves (*Beta vulgaris*) in a completely controlled environment. Figure 11 shows the test results with sugarbeet leaves.

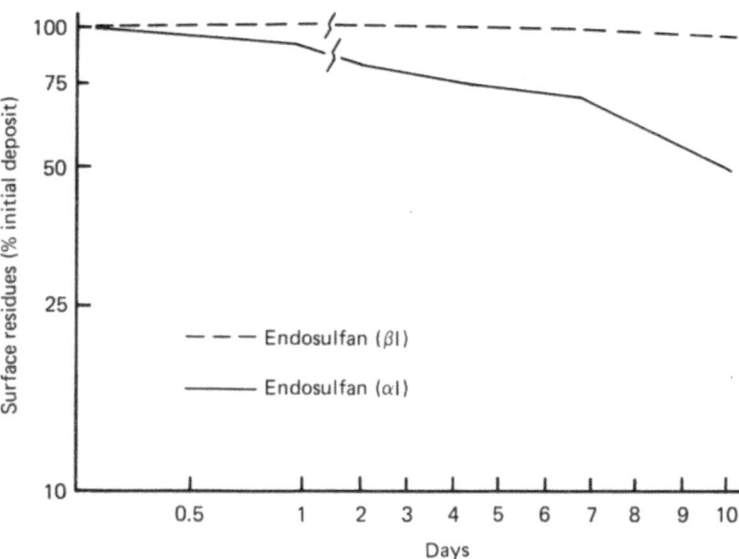

Fig. 10. Volatility of endosulfan (αI and (βI) (15.7 μg/cm²) from glass plates (Beard and Ware 1969).

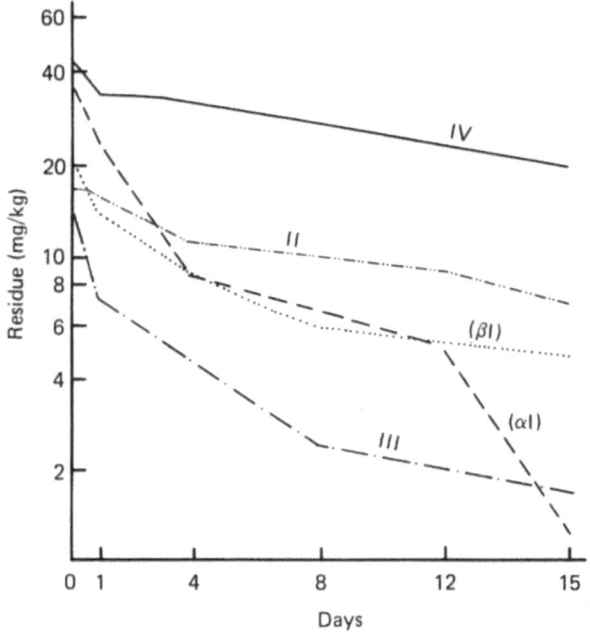

Fig. 11. Decrease of endosulfan (αI) and (βI), endosulfan diol (II), endosulfan sulfate (IV), and endosulfan ether (III) on sugarbeet leaves under controlled conditions (Beard and Ware 1969).

The said volatility properties account for the composition of terminal residues, as they occur after treatment with endosulfan (I). Immediately after treatment, the isomers in the residue are in a ratio of 2:1. This ratio changes fundamentally after approximately 14 days' weather exposure. The ratio of isomers is shifted towards endosulfan (βI). Endosulfan sulfate (IV), which evolves as metabolite and is less volatile than both endosulfan (αI) and (βI), may constitute up to 50% and more of the total residue.

2. **Adsorption.**–Model tests with iron hydroxide gel, activated carbon, and river silt were reported by Greve (1971). Figures 12 and 13 show as examples two adsorption isotherms (Freundlich's adsorption isotherms).

Iron hydroxide gel exerts a catalytic effect on the hydrolysis rate of endosulfan (I) to endosulfan diol (II).

3. **Hydrolysis.**–Hydrolysis tests in sterile water are not known. Greve (1971) reported on tests in water at pH 7 and pH 5.5 under aerobic and anaerobic conditions. According to said tests, hydrolysis proceeds much faster at pH 7 than at pH 5.5. In terms of half-life: at pH 7.0 one-half is hydrolyzed after approximately 5 wk, at pH 5.5 only after 5 mon. Similar results were reported by Schöttger (1970). Hydrolysis is considerably accelerated by mixing iron hydroxide gel to the water. Especially the endosulfan (I) adsorbed by the iron hydroxide gel is subject to very rapid degradation to endosulfan diol (II) which, owing to its LD_{50} value in rats of > 15,000 mg/kg, can be considered toxicologically unobjectionable (Greve 1971). This finding is, therefore, of importance for evaluating the environmental behavior of endosulfan.

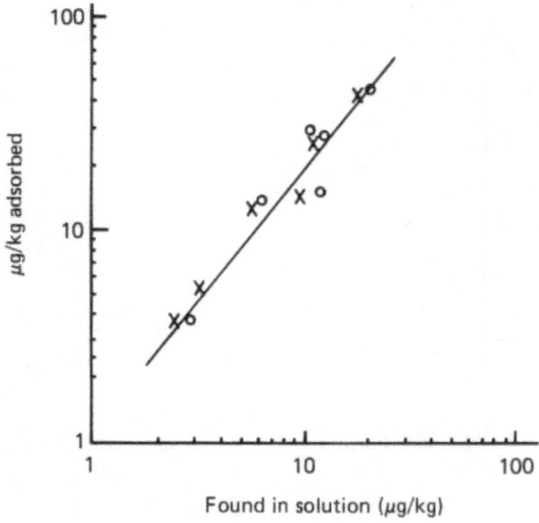

Fig. 12. Freundlich's adsorption of endosulfan (I) on iron hydroxide (Greve 1971); 50 mg of $Fe(OH)_3$-gel/kg of silt.

The rate of hydrolysis is likewise increased considerably in water, if the latter contains a major amount of algae and/or bacterial sludge. Gorbach and Knauf (1971) showed that endosulfan (I) could be added daily in a rhythm of 10 mg/day/L to water to which bacterial sludge from a biological sewage purification plant had been added (0.5 g dry substance/L of water). The balance at the end of the seventh day revealed that approximately 50% of the endosulfan (I) passed through had been degraded to endosulfan diol (II).

A similar rapid hydrolysis to endosulfan diol (II) is observed if the water contains green algae (*Chlorella, Scenedesmus*) (Gorbach and Knauf 1971).

It was already mentioned in Section III that endosulfan (I) can be transformed to endosulfan diol (II) in warm-blooded animals and by the action of soil microorganisms. Especially in the presence of organisms that take part in the conversion of endosulfan (I) to endosulfan diol (II), this latter compound can constitute the terminal product of a gradual degradation.

Insufficient knowledge exists on the evolution of endosulfan diol (II) as residue on plants following endosulfan (I)-treatment. A major test series was reported by Archer (1973), for which, however, an endosulfan (I) had been used that already contained 8.9% endosulfan diol (II) and did not correspond with the technical grade concerning the ratio of isomers. From this test result the author came to the conclusion that a slight increase of endosulfan diol (II) takes place under the influence of sunlight. This finding is in agreement with the test results of Archer *et al.* (1972), who showed that under irradiation and free access of atmospheric air, predominantly endosulfan diol (II) evolves from endosulfan (I).

4. Oxidation.—The oxidation of endosulfan (I) yields endosulfan sulfate (IV), which is an essential residue constituent after endosulfan application.

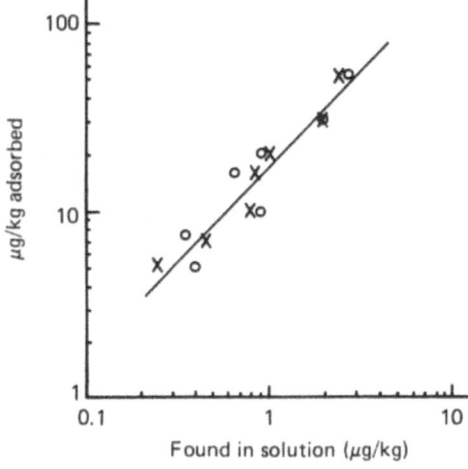

Fig. 13. Freundlich's adsorption of endosulfan (I) on river silt I, 1% (Greve 1971).

The papers published so far do not give information beyond any doubt as to the mode of evolution under natural environmental atmospheric conditions.

Harrison et al. (1967) as well as already Cassil and Drummond (1965) ascertained that endosulfan (I) is not only likely to be converted by UV irradiation in the air on an inert surface to form endosulfan sulfate (IV). It was only after addition of (not further identified) enzymes extracted from apple leaves, that Harrison et al. (1967) observed a minor formation of endosulfan sulfate (IV). Without light, however, the enzyme system failed to accomplish oxidation!

An experiment with lucerne, carried out by Archer (1973) stands in contrast thereto. He found that the formation of endosulfan sulfate (IV) was an optimum if the endosulfan (I)-treated lucerne cured in the dark (cp., Fig. 3 in Section III). This finding was confirmed by Gorbach and Bock (1974) in a still unpublished test series.

5. Photodegradation.—The essential reactions proceeding under the influence of light of different wavelengths have already been described in paragraph 1.

6. Biochemical degradation.—The various metabolites biochemically formed from endosulfan (I) in the live organism were discussed extensively in Section III. A few remarks might be added which may specifically refer to certain crops. This applies primarily to the varying amount of endosulfan sulfate (IV) evolved, which is undesirable as residue. The residue figures in Tables XVII to XXIV teach that the endosulfan sulfate (IV) proportion is rather variable in the terminal residue. From the results at hand it cannot be concluded definitely whether the endosulfan sulfate (IV) has chiefly evolved during the curing of the leaves, as in the experiments by Archer (1973) with lucerne.

b) Endosulfan residues in the soil

The degradation behavior of endosulfan in laboratory tests and the degradation products obtained have already been discussed in Section III. It should be added here, what kind of residues would have to be anticipated in the soil after application of endosulfan in the field.

Results from extensive investigations on residues are the subject of still unpublished reports on industrial operations.

In 1964, *Niagara Chemical Division of FMC* conducted in two places, Jackson, MS and Gasport, NY, controlled residue tests, using in one experiment the conventional dose of 2 lb/A, and in a second experiment ten-fold said dose. The trials in Gasport were extended in a manner such that the active substance was on the one hand not incorporated, and on the other hand incorporated into the soil.

The results of said experiments are presented in Tables XVII to XXII.

Some further results of soil samples from several other locations in the U.S.A. are given in Table XXIII. Soil samples from experimental plots in Modena, Italy, that are used for sugarbeet cultivation every year and that were partly treated in more than one vegetation period with Thiodan, were examined by *Hoechst AG* (1973) for endosulfan residues.

The results are compiled in Table XXIV. Gorbach *et al.* (1971 b) reported on the degradation in the soil of paddy fields.

Within the scope of a permanent control of plant pesticide residues in the soils of the U.S.A. soil samples from 51 locations of the U.S.A. were analyzed by Stevens *et al.* (1970) in the years 1965, 1966, and 1967. Endosulfan was detected in only a few cases, although it had been used in the respective areas.

In a similar investigation in 1967 and 1968, Mullins *et al.* (1971) found endosulfan (I) only in a single case in soil samples from fruit cultures and other crop areas in Colorado (U.S.A.).

Foschi *et al.* (1970) compared the persistence of a series of plant pesticides, including endosulfan (I), and classified endosulfan (I) as less persistent than, *e.g.*, DDT, dieldrin, Temik, Captan, and Benlate, but as slower in degradation as compared with most of the organophosphorous esters.

A study of several years' duration (1964, 1966, and 1969) dealing with the residue pattern in the soil of 16 farms in the southwest of the Province of Ontario (Canada) was published by Harris and Sans (1971). Also, in this case endosulfan (I) was detected in isolated samples only.

Miles and May (1978) reported the results of a further monitoring program (1972 to 1975) on insecticide residues in organic soils of Holland Marsh (Ontario, Canada).

Within the scope of the National Soils Monitoring Program, 1970, Crochett *et al.* (1974) examined soil samples from 1,500 cultivation areas in 35 states in the U.S.A. for plant pesticide residues, including endosulfan (I). This study also revealed that endosulfan could be detected in but a few cases, and this only in soils of treated cultures.

In similar investigations carried out by Carey (1979), endosulfan has not been traced as a soil residue.

Of course the degradation of endosulfan (I) depends on the one part on the nature of the soil, *viz.*, its chemical composition, and on the other part—not the least—on the quality and quantity of the biomass (fungi, bacteria), as was shown in Section III, *a*). A relative short persistence in soils (100 to 120 days) compared to the chlorinated pesticides was stated very recently by Rao and Murty (1980).

c) Endosulfan (I) residues in waters

In discussing the factors that have an influence on residues, this section has already dealt with the degradation pattern of endosulfan (I) in water.

The behavior of endosulfan (I) in surface waters, starting from reported results, is quite variable and rather dependent on the animation and the pH value of the water. Further, the concentration of certain suspended particles is of great influence on the persistence of residues in the water.

Eichelberger and Lichtenberg (1971) have reported on comparative degradation tests with several plant pesticides in crude water of the Little Miami River (U.S.A.) carrying a medium heavy load of house and industrial sewage as well as

Table XVII. *Endosulfan residues in soil samples (after Niagara Chem. Div. 1964).*[a]

Preparation	Formulation	Kind of sample	Location	No. of applications	Last application	Sampling date	a.i. (lb/A)	Days after last treatment	Residues (mg/kg)[b]
Thiodan	2 EC	Soil (0-6 in.) incorporated	Niagara Research Farm, Jackson, MS	1	6/12/64	6/12/64	2.0	0	1.8
								38	0.70
								62	0.80
								96	0.60
								130	0.24
								276	0.75
								307	0.81
								336	0.35
								368	0.38
								398	0.33
								460	0.08
								503	0.27
Thiodan	2 EC	Soil (6-12 in.) not incorporated	Niagara Research Farm, Jackson, MS	1	6/12/64	6/12/64	2.0	0	0.34
								38	0.86
								62	< 0.05
								96	n.d.
								103	n.d.
								276	n.d.
								307	n.d.
								336	n.d.
								368	< 0.05
								398	< 0.05
								460	< 0.05
								503	n.d.

Thiodan	2 EC	Soil (12-18 in.) not incorporated	Niagara Research Farm, Jackson, MS	1	6/12/64	6/12/64	2.0	0	0.06
								38	<0.05
								62	<0.05
								96	n.d.
								130	n.d.
								276	n.d.
								307	n.d.
								336	n.d.
								368	<0.05
								398	n.d.
								460	n.d.
								503	n.d.

[a] = Analytical method, MCGC.
[b] n.d. = not detectable.

Table XVIII. *Endosulfan residues in soil samples (after Niagara Chem. Div. 1964).* [a]

Preparation	Formulation	Kind of sample	Location	No. of applications	Last application	Sampling date	a.i. (lb/A)	Days after last treatment	Residues (mg/kg) [b]
Thiodan	2 EC	Soil (0-6 in.) incorporated	Niagara Research Farm, Gasport, NY	1	6/22/64	6/22/64	2.0	0	1.6
								29	1.6
								59	0.73
								91	1.0
								122	1.0
								349	0.65
								409	0.36
								469	0.25
Thiodan	2 EC	Soil (6-12 in.) not incorporated	Niagara Research Farm, Gasport, NY	1	6/22/64	6/22/64	2.0	0	0.36
								29	0.50
								59	0.34
								91	0.41
								122	0.41
								349	< 0.05
								409	0.07
								469	0.21
Thiodan	2 EC	Soil (12-18 in.) not incorporated	Niagara Research Farm, Gasport, NY	1	6/22/64	6/22/64	2.0	0	0.26
								29	0.21
								59	< 0.05
								91	0.12
								122	n.d.
								349	n.d.
								409	n.d.
								469	n.d.

[a] = Analytical method, MCGC.
[b] n.d. = not detectable.

Table XIX. *Endosulfan residues in soil samples (after Niagara Chem. Div. 1964).*[a]

Preparation	Formulation	Kind of sample	Location	No. of applications	Last application	Sampling date	a.i. (lb/A)	Days after last treatment	Residues (mg/kg)
Thiodan	2 EC	Soil (0-6 in.) not incorporated	Niagara Research Farm, Gasport, NY	1	6/23/64	6/23/64	2.0	0	1.0
								28	0.90
								58	1.1
								90	1.1
								121	0.68
								349	0.64
								409	0.05
								469	0.23
Thiodan	2 EC	Soil (6-12 in.) not incorporated	Niagara Research Farm, Gasport, NY	1	6/23/64	6/23/64	2.0	0	0.98
								28	0.57
								58	0.18
								90	0.53
								121	0.16
								349	0.36
								409	< 0.05
								469	0.20
Thiodan	2 EC	Soil (12-18 in.) not incorporated	Niagara Research Farm, Gasport, NY	1	6/23/64	6/23/64	2.0	0	0.43
								28	0.80
								58	0.29
								90	0.20
								121	0.22
								349	0.05
								409	0.10
								469	0.11

[a] Analytical method, MCGC.

Table XX. *Endosulfan residues in soil samples (after Niagara Chem. Div. 1964).*[a]

Preparation	Formulation	Kind of sample	Location	No. of applications	Last application	Sampling date	a.i. (lb/A)	Days after last treatment	Residues (mg/kg)
Thiodan	2 EC	Soil (0-6 in.) not incorporated	Niagara Research Farm, Jackson, MS	1	6/12/64	6/12/64	20.0	0	16.7
								38	9.5
								62	7.6
								96	6.4
								130	3.6
								276	5.9
								307	5.3
								336	5.4
								368	4.4
								398	2.3
								460	1.3
								503	1.5
								530	3.4
Thiodan	2 EC	Soil (6-12 in.) not incorporated	Niagara Research Farm, Jackson, MS	1	6/12/64	6/12/64	20.0	0	4.0
								38	0.21
								62	0.80
								96	0.25
								130	0.08
								276	< 0.05
								307	0.15
								336	0.28
								368	0.64
								398	0.06
								460	0.19

								503	0.16
								530	0.35
Thiodan	2 EC	Soil (12-18 in.) not incorporated	Niagara Research Farm, Jackson, MS	1	6/12/64	6/12/64	20.0	0	1.6
								38	0.68
								62	0.41
								96	0.31
								130	0.14
								276	< 0.05
								307	0.12
								336	< 0.05
								368	< 0.05
								398	< 0.05
								460	0.13
								503	0.12
								530	0.28

[a] Analytical method, MCGC.

Table XXI. *Endosulfan residues in soil samples (after Niagara Chem. Div. 1964).*[a]

Preparation	Formulation	Kind of sample	Location	No. of applications	Last application	Sampling date	a.i. (lb/A)	Days after last treatment	Residues (mg/kg)
Thiodan	2 EC	Soil (0-6 in.) incorporated	Niagara Research Farm, Gasport, NY	1	6/22/64	6/22/64	20.0	0	11.7
								29	6.2
								91	5.8
								122	2.3
								349	6.2
								409	2.1
								469	0.91
Thiodan	2 EC	Soil (6-12 in.) not incorporated	Niagara Research Farm, Gasport, NY	1	6/22/64	6/22/64	20.0	0	0.91
								29	0.94
								91	3.0
								122	2.3
								349	1.6
								409	0.71
								469	1.7
Thiodan	2 EC	Soil (12-18 in.) not incorporated	Niagara Research Farm, Gasport, NY	1	6/22/64	6/22/64	20.0	0	0.89
								29	1.5
								91	1.1
								122	1.0
								349	0.66
								409	0.55
								469	0.49

[a] Analytical method, MCGC.

Table XXII. *Endosulfan residues in soil samples* (after *Niagara Chem. Div.* 1964).[a]

Preparation	Formulation	Kind of sample	Location	No. of applications	Last application	Sampling date	a.i. (lb/A)	Days after last treat-ment	Residues (mg/kg)
Thiodan	2 EC	Soil (0-6 in.) not incorporated	Niagara Research Farm, Gasport, NY	1	6/23/64	6/23/64	20.0	0	8.6
								28	6.8
								58	5.9
								121	8.0
								349	–
								409	4.2
								469	1.6
Thiodan	2 EC	Soil (6-12 in.) not incorporated	Niagara Research Farm, Gasport, NY	1	6/23/64	6/23/64	20.0	0	2.6
								28	6.0
								58	1.8
								121	0.83
								349	1.1
								409	0.40
								469	0.78
Thiodan	2 EC	Soil (12-18 in.) not incorporated	Niagara Research Farm, Gasport, NY	1	6/23/64	6/23/64	20.0	0	1.1
								28	1.2
								58	1.0
								121	0.22
								249	0.97
								409	0.75
								469	0.72

[a] Analytical method, MCGC.

Table XXIII. *Endosulfan residues in soil samples (after Niagara Chem. Div. 1964).*

Preparation	Formulation	Kind of sample	Location	No. of applications	Last application	Sampling date	a.i. (lb/A)	Days after last treatment	Residues (mg/kg)	Analysis method
Thiodan	50 WP + 2 EC	sandy loam (tomato + paprika field)	Nickelton, NJ	30 from 6/10/65	8/15/69	6/2/70	1.0	293	0.34	MCGC
Thiodan	—	sandy loam (potato field)	Pickerel, WI	7 from July 66	8/9/69	5/12/70	0.5	282	0.41	
Thiodan	Staub 3%	sandy loam (tobacco field)	Quincy, FL	9 from 5/6/69	7/1/69	6/3/70	0.5-0.6	336	1.2	
Thiodan	—	muddy, clayey loam (potato + cabbage field)	Homestead, FL	10	1970	6/22/70	1.0	ca. 100	0.23	
Thiodan	—	clayey loam (potato + cabbage field)	Goulds, FL	15-17 from 1966	1969	6/22/70	1.0	ca. 100	0.14	
Thiodan	2 EC	soil	NCD Research Farm, Davis, CA	1	7/29/71	7/29/71	1.0	0	0.27	GLC-EGD
								3	0.19	
								10	0.16	
								17	0.13	
								24	0.13	
								31	0.08	

Table XXIV. *Residues of endosulfan (I) and endosulfan sulfate (IV) in soil samples from sugarbeet field near Modena (Italy) (Hoechst AG 1973-1974).*

Application time, each 0.5 kg/ha a.i.	Sampling date	Residue [mg/kg] in soil depth (cm)[a]					
		(0 - 5)		(5 - 10)		(10 - 15)	
		(I)	(IV)	(I)	(IV)	(I)	(IV)
1969 2 x Apr., June	Sept. 70	0.1	0.2	0.05	0.04	0.05	0.04
1970 Apr., July, Aug.	Mar. 71	0.024	0.02	0.04	0.04	0.05	0.04
1971 May, June, July	Sept. 71	0.01	0.16	n.d.	0.01	n.d.	0.009
	Mar. 72	n.d.	n.d.	n.d.	0.002	n.d.	0.006
1972 Apr., June, July	Oct. 72	0.04	0.09	0.001	0.007	0.002	0.02
	Mar. 73	0.009	0.016	0.015	0.024	0.009	0.02
1973 May, June, July	Sept. 73	0.045	0.15	0.009	0.019	0.008	0.007
	Apr. 74	0.025	0.08	0.026	0.1	0.015	0.06

[a] n.d. = not detectable.

an afflux of farmland seepage water. The insecticides were added in a concentration of 10 μg/L of water. As the following abbreviated Table XXV reveals, endosulfan has a very low persistence in the water examined, according to these results a circumstance very likely attributable to the relatively high pH value of 7 to 8 (cp., Greve 1971).

Gorbach et al. (1971 b) described a degradation test in a paddy field in East Java. The degradation graph after application of 1.5 L Thiodan 35 EC/ha in a single dose is shown in Figure 14.

Gorbach et al. (1971 a) described a study on endosulfan (I) residues in the water system of an extended rice cultivation area in East Java after extensive application of endosulfan. Approximately 400 tons of endosulfan (I) (formulated as EC) were applied on a cultivation area of approximately 133,000 ha in the catchment area of the river Brantas, having a length of 200 km (Jan. 24%, Feb. 33%, Mar. 24%, Apr. 9.5%). The investigations began shortly after the application of the bulk of the material in March. Table XXVI presents the results obtained.

Several studies were carried out in order to detect the amount of endosulfan (I) in the water run off from the treated cultivation areas.

Epstein and Grant (1968) examined the drainage water from a loam region (Presque Isle, ME). An extended surface had been previously sprayed with 0.35 kg/ha of endosulfan. Four days after the application and after a heavy rainfall, a mean content of 16 μg/L was recovered in the run-off water. Within 12 days during which further heavy rainfalls took place, the content in the run-off water at the same sampling location dropped to an average of 1 μg/L.

Miles and Harris (1971) measured the endosulfan (I) concentration in drainage ditches of a farming area near Lake Erie at Erieau, Ontario. In the soil of a farm situated not far from the drainage ditch, there was found 0.640 mg/kg

Table XXV. *The persistence of some pesticides in river waters[a]* (after Eichelberger and Lichtenberg 1971).

Active substance	Unchanged active substance, recovered in % of addition[b]				
	Time in weeks				
	0	1	2	4	8
Dieldrin	100	100	100	100	100
Endosulfan	100	30	5	n.d.	n.d.
Parathion	100	50	30	5	n.d.
Dimethoate	100	100	85	75	50
Sevin (carbaryl)	90	5	n.d.	n.d.	n.d.
Monuron	80	40	30	20	n.d.

[a] pH value of the water 7-8, hardness 150 mg/L, phosphorus 0.73 mg/L, ammonia plus organic nitrogen 4.3 mg/L, carbon 10.7 mg/L (total), and nitrate 1.27 mg/L.

[b] n.d. = not detectable.

of endosulfan. During the whole sampling period, the endosulfan content in the ditch-water varied between 0.002 and 0.18 µg/L. The mud in the ditch had residues between 0.001 and 0.06 mg/kg.

A widespread dying of fish of the Rhine, beginning in the Geisenheim section on June 23, 1969, suggested an accidental contamination of the water by endosulfan (I). The water analyses in the period of June 19, 1969, to November 8, 1969, revealed a mean concentration of 0.1 to 0.3 µg/L. After decrease of the concentration to approximately 0.07 µg/L, there appeared in November a new maximum of 0.49 µg/L. No dying of fish was observed at this time, although the Rhine now had a stock of fish again in the respective sections of the river. The concentrations were under the lethal concentration threshold of 1 µg/L for the most susceptible variety of fish (upper toxic limit for fish 10 µg/L). Since a simultaneous influence of several toxicants could not be excluded in the polluted Rhine water, the true cause for the death of the fish could not be clarified (*Hoechst AG* 1970). Further results concerning this casualty were published by Lüssem and Schlimme (1971), as well as by Greve and Wit (1971).

Further investigations carried out later on river pollutions in the Federal Republic of Germany, particularly concerning the lower course of the Main River, were published by Herzel (1972).

Table XXVII shows results of current controls of the lower course of the Main River near Kostheim (27 to 28 river km off the plant site) carried out by *Hoechst AG* (unpublished).

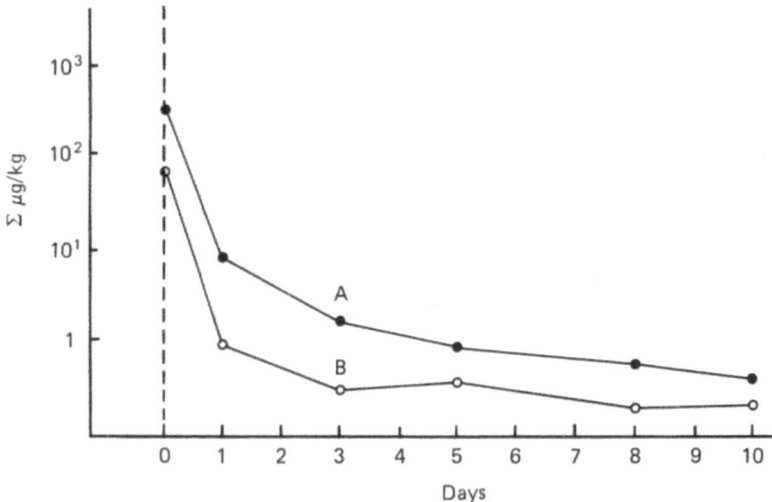

Fig. 14. Decrease of endosulfan in the water of a constantly watered paddy field (after Gorbach *et al.* 1971 b). pH value of applied water 6.5, influx of fresh water 1.5 L/sec, size of field A = 64 m², B = 173 m², endosulfan rate applied 1.5 L Thiodan® 35 EC/ha.

Table XXVI. *Residues of endosulfan (I) and endosulfan sulfate (IV) in µg/L in the water system of the Brantas River (East Java), the canals and fish ponds in the Brantas delta, and in the Madura Sea* (after Gorbach *et al.* 1971 a).

Location	Mean values and range of values in µg/L[a]		
	Endosulfan		Endosulfan sulfate (IV)
	αI	βI	
Brantas River	< 0.1	< 0.1	< 0.19
	0.01-5.0	< 0.01-2.0	< 0.01-0.45
Canals	< 0.13	< 0.12	< 0.18
	< 0.01-5.8	< 0.01-2.4	< 0.01-0.55
Fish ponds	< 0.03	< 0.02	< 0.06
	< 0.01-0.25	< 0.01-0.08	< 0.01-0.44
Madura Sea	< 0.02	< 0.02	< 0.08
(coastal waters)	< 0.01-0.09	< 0.01-0.07	< 0.01-0.28
Madura Sea	n.d. ≅ < 0.01	n.d. ≅ < 0.01	n.d. ≅ < 0.01

[a] n.d. = not detectable.

Bauer (1972) reported on the elimination of endosulfan residues from the water. By treatment with 5 mg of permanganate/L, water residues of 40 mg/kg could be completely eliminated.

Results of several years' control (1969 to 1975) of water samples from Dutch waters were published by Wegmann and Greve (1978).

In summary it may be stated that the residue values of endosulfan (I) in water decrease continuously in a relatively short time after contamination to unobjectionable values or to values below the limit of detection.

Table XXVII. *Medium and maximum concentrations of endosulfan in the water of the Main River near Mainz-Kostheim (Hoechst AG, unpublished).*

Year	No. of yearly samplings	Residue (µg/L)[a]	
		Mean value	Maximum value
1970	12	1.5	3
1971	20	0.3	0.9
1972	10	n.d.	—
1973	22	n.d.	0.3
1974	23	n.d.	0.3

[a] n.d. = not detectable.

d) Uptake of endosulfan residues by animals and plants

1. Water as environment.—A review of this subject was given by the *National Research Council* (1975). This critical assessment led to the opinion that endosulfan (I) and endosulfan sulfate (IV), in contrast to the chlorinated hydrocarbons, do not show a tendency in the ecosystem water to accumulate in the lipid-rich pools.

This assertion is based on a series of papers in which, especially, the uptake and excretion of endosulfan by fish was investigated. A first very detailed paper was published by Schoettger (1970). Western white suckers (*Catostomus commersoni*) had been exposed for 12 hr to water containing 20 μg of endosulfan (I)-^{14}C/L. It was found that in virtually all examined tissues the radioactivity had reached a plateau within less than 12 hr. The concentrations reached after 9 and 12 hr in the respective tissues and the corresponding concentration factors are compiled in Table XXVIII.

A constancy of the plateau over a prolonged period of time could be revealed in experiments with goldfish (*Carassius auratus*), that were repeatedly treated with daily fresh endosulfan (I) solutions (7 μg/L). The residue value in the muscle flesh, determined after a test period of 5 days, was 2.5 mg/kg, and after a 20-day test period it was 1.9 mg/kg.

In experiments reported by Oeser *et al.* (1971) it could be proved that the endosulfan taken up by goldfish (*Carassius auratus*) was excreted with a half-life period of 2 to 3 days after transfer of the fish to endosulfan-free water. To this effect, each of 5 groups of fish were exposed to an endosulfan (I) concentration of 1 μg/L in 5 aquariums. Every day 90% of the water was replaced by fresh

Table XXVIII. *Concentrations of radioactive substance in several tissues of Catostomus commersoni (western white suckers) following exposure of 20 μg/L of* [14]*C-labeled endosulfan (I) at 19°C* (after Schoettger 1970).

Tissue	Residues of radioactive substance (μg/g dry tissue)		Concentration factor[a]	
	Exposure time		Exposure time	
	9 hr	12 hr	9 hr	12 hr
Liver	13.9	11.1	695	550
Kidney	2.3	4.4	115	220
Gut	10.9	3.0	545	150
Skin	1.3	4.9	65	245
Muscle	1.1	1.3	55	65
Gill	3.2	6.4	160	320

[a] Concentration factor $= \dfrac{\text{μg/g endosulfan in resp. tissue}}{\text{μg/g endosulfan in water}}$

water of the same endosulfan concentration, thus warranting a virtually constant concentration of 1 μg/L. After a lapse of 5 days, the fish of the first group were examined for any residue and the remaining groups transferred further on separately into fresh water free from any active substance. After 1, 3, 7, and 14 days, these groups were examined for residues. The results showed that the fish had taken up on the average 0.35 mg/kg active substance (referred to wet wt). Residues on the skin, in relation to the total wt of the fish, were practically negligible (0.002 mg/kg). The rate of excretion of the residues after transfer to fresh water is demonstrated in Figure 15.

Autoradiograms of fish (*Lebistes reticulatus*, guppy) (Knauf *et al.* 1976) showed that already one day after putting the fish into the container stocked with fish-feed organisms (*Daphnia magna*, water-flea) and provided with ^{14}C-endosulfan (I), radioactive residues became visible first in the liver. The other organs, only very much later, showed residues in minor amounts as compared with the liver. After transfer of the fish into fresh water, the radioactive residues were likewise excreted very rapidly.

Residue examinations in fish that had lived in endosulfan-contaminated natural surface waters and that were caught in a fishery manner showed a mean residue value of 0.4 mg/kg (in terms of fresh fish substance) (Gorbach and Knauf 1971). Roberts (1972) kept mussels (*Mytilus edulis*) in seawater with 0.1, 0.5, and 1 mg endosulfan (I)/L for 112 days. The highest residue count determined after this period was approximately 8 mg/kg, and it dropped within 58 days, during which the mussels lived in fresh water, to 2.5 mg/kg.

In a later paper Roberts (1975) investigated the route by which the endosulfan residues got into the mussels. His findings suggested that especially the oral intake of particles with endosulfan adsorbed thereto appeared to be responsible

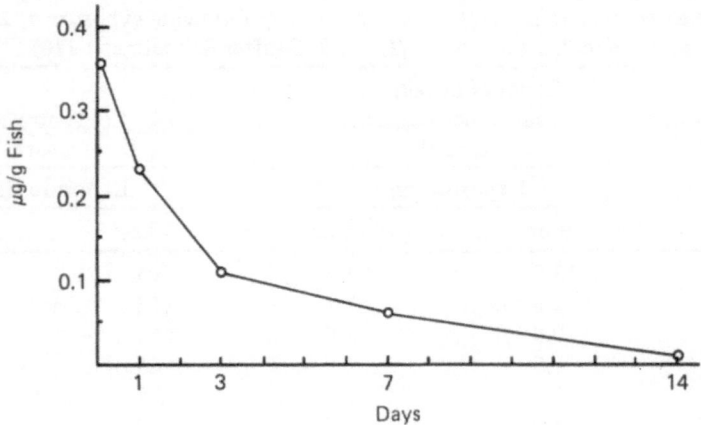

Fig. 15. Decrease of residue (Σ endosulfan (I) and endosulfan sulfate (IV)) after setting residue-containing goldfish (*Carassius auratus*) into endosulfan-free water.

therefor, since the digestive gland with 3.8 $\mu g/g$ had the highest residue count, whereas the gills contained only 0.46 $\mu g/g$.

An ecological study is of interest which was carried out by Koeman *et al.* (1974) in Java (*cp.*, Table XXIX). Besides endrin, endosulfan (I) was essentially used in intensive, several years' so-called BIMAS (mass guidance) programs for the control of the paddy stem borer (*cp.*, also Gorbach *et al.* 1971 b).

Koeman *et al.* (1974) studied the animal varieties given in Table XXIX and the eggs of sasah ducks. The results of these costly studies show that endosulfan (I) and its metabolite endosulfan sulfate (IV) exhibits no residues in the animal varieties examined in Java (detection limit 0.03 mg/kg).

Aquatic plants, especially green algae, are capable of absorbing endosulfan within a few hr. In this connection, a concentration factor of approximately 2,000 was observed. The absorbed endosulfan is slowly metabolized to endosulfan diol (II), which was given off into the surrounding water, where it was detected (Gorbach and Knauf 1971, Oeser *et al.* 1971).

Frank *et al.* (1978) reported on current supervision (1968 to 1976) of the fish population in Lake Saint Clair and Lake Erie, Canada, with regard to organochlorine plant pesticides, including endosulfan.

In summary it can be assumed that endosulfan (I) is eliminated in water within a relatively short time, *i.e.*, that a long-term influence on the ecosystem especially with regard to the fauna fish feed on is not to be feared. This can be concluded from the rapid regeneration of the stock of fish in natural waters.

Table XXIX. *Description of samples analyzed for endosulfan residues (Origin. Java)* (Koeman *et al.* 1974).

	Species	No. of samples analyzed
Fish:	*Chanos chanos* from tambaks	6
		6
	Cyprinus carpio	3
		3
		4
	Hampala macrolepidota	7
	Leiognathus splendens	50
	Macrones nemurus (fat)	1
	Mystacoleucus marginatus	10
	Pangasius pangasius	5
	Saurida sp.	10
	Tilapia mossambica	1
	Trichogaster pectoralis	10
	Upeneus sulphureus	15
Molluscs:	various sp. of bivalves	102
Crabs:	*Paratelphusa* sp.	16
Shrimps:	(local name *Udang apik*)	55

2. Terrestrial environment.—Endosulfan (I) orally taken up by warm-blooded animals, even in high doses, as such does not invade the meat, fat, or the milk. However, depending on the administered dose, the metabolite endosulfan sulfate (IV) is detected in the fatty tissue and in the milk (Gorbach *et al.* 1968, McCaskey and Liska 1967, Beck *et al.* 1966, and survey by Maier-Bode 1968; further Schmidlin-Mészáros and Romann 1971). These last authors described a case of cattle intoxication due to feeding on contaminated hay, from which a daily dose of endosulfan (I) amounting to 10 to 20 mg/kg of cattle body wt could be computed. The analyses of the milk and liver yielded only traces (< 14 μg/kg) of endosulfan, although the cattle exhibited distinct symptoms of intoxication. Endosulfan is excreted (as given in Section III more in detail) partly unchanged with the feces, partly in the form of water-soluble metabolites with the urine (Maier-Bode 1968, Elzner 1973).

From the data at hand it is doubtless that the concentration factor for endosulfan lies below 10^{-3} when it passes from the fodder into the fatty tissue [chlorinated hydrocarbons such as DDT, dieldrin, etc., possess a concentration factor of 12 (Bressau 1975)].

Reference is also made to the survey published by the *National Research Council Canada* (1975).

The endosulfan sulfate (IV) residues in tissue rather rapidly reach a plateau and they are—after discontinuation of endosulfan (I) administration—quickly eliminated (Kloss *et al.* 1966).

The uptake of endosulfan by insects is described in a summary by Maier-Bode (1968). It is clearly shown that endosulfan (I) does not give rise to persistent residues in the animal body. Uk and Himel (1972) found a half-life value of 83 min for endosulfan residues on the surface of houseflies, and of 176 min for the total body.

3. Laboratory ecosystem.—The behavior of endosulfan (I) in a laboratory ecosystem was described by Ali (1978). The partial conversion of endosulfan (αI) to endosulfan sulfate (IV) by all animals living in the ecosystem (fish, snails, etc.) was observed. Residues were recorded particularly in fish and snails. The mouse was also comprised in the study, and the metabolites observed were in accordance with those described in Section III. In the organs and in the fat, however, no residues were detected. A rapid degradation of endosulfan (I) in the water under the influence of light was reported.

e) Residues in food and stimulants

1. Vegetable foodstuffs and forage crops.—Maier-Bode (1968) gave a survey of the results obtained until 1967 in this periodical.

Since then, a greater number of new results has been found which were only partly published. The survey hereinafter given and Table XXX show a compilation of the data that have come to hand. The residues reported therein were evidenced after conventional sprayings in the respective crops. The results were obtained

Table **XXX**. *Endosulfan residues in different crops* (*Hoechst AG*, unpublished).

Crop	No. of applications	kg a.i./ha	Days / weeks after last application		Residues[a] (mg/kg)
Sweet corn kernels	6	1.68	0		n.d.
Sweet corn cobs	6	1.68	0		n.d.
Sugarbeet root	3	1.12	0		n.d.
Sugarbeet pulp	3	1.12	0		n.d.
Onion hollow	1	0.21	1		n.d.
Onion string	1	0.21	1		n.d.
Oil palm	1	0.525	30		0.28
Oil palm	1	0.35	30		0.2
Rape seed	1	0.84		11	n.d.
Rape seed	1	0.42		11	n.d.
Clover	1	0.21	7		1.35
Alfalfa	1	0.21	7		0.14

[a] n.d. = not detectable.

from numerous degradation series and permit, therefore, an estimation of the residues occurring in practical application.

α).*Apple*.–At the application rate of 0.95 kg a.i./ha the residues remain 1 to 2 wk after the last application below the MRL and range between 0.84 and 0.25 mg/kg, and at 0.73 kg a.i./ha, the residues are 0.27 to 0.46 mg/kg 1 wk after the last application (MRL effective for the Federal Republic of Germany, U.S.A., FAO/WHO; *cp.*, also Section VIII).

β).*Grape*.–At the rate of 2.8 kg a.i./ha, the residues remain 2 wk after the last application below the MRL (*cp.*, Section VIII), at 1.68 kg a.i./ha already so after 1 wk. They are at 0.9 to 0.17 mg/kg. Singh and Chawala (1979) have come to a similar evaluation of the results.

γ).*Cabbage*.–At the rate of 0.525 kg a.i./ha, the residues are 4 wk after the last application below the MRLs (cp., Section VIII), and at 0.35 kg a.i./ha, they are so after 2 wk.

δ).*Tomato*.–At the rate of 1.2 kg a.i./ha (California), the residues are found to be below the MRLs (*cp.*, Section VIII) already 1 wk after the last application. In the Federal Republic of Germany (Hesse), at the rates of 0.525 and 0.21 kg a.i./ha, the residues lie 1 wk after the last application below the MRLs.

ε).*Spinach*.–At the rate of 0.21 kg a.i./ha, the residues remain below the MRLs already 5 days after the last application (*cp.*, Section VIII). They lie between 0.2 and 1.9 mg/kg, and 10 days after the last application between 0.1 and 0.5 mg/kg.

ς).*Lettuce.*—At the application rate of 0.21 kg a.i./ha, the residues were below the MRLs already 5 days after the last application (*cp.*, Section VIII). They were between 0.17 and 0.95 mg/kg.

η).*Cotton.*—At 1.12 kg a.i./ha, the following residues were found 9 days after application:

ginned seed	0.19 mg/kg
linters	0.26 mg/kg
linters mote	0.33 mg/kg
hulls	0.01 mg/kg
meal	0.01 mg/kg
crude oil	0.15 mg/kg
refined oil	n.d.
soap stock	0.02 mg/kg

ϑ). *Sugarcane.*—After treatment of sugarcane with endosulfan (I), Awasthi *et al.* (1977) did not find any residue in the pressed syrup.

ι). *Sweet potatoes.*—After treatment of sweet potatoes with endosulfan (I), Rajukkannu *et al.* (1978) did not find any residues beyond the MRL (< 0.5 mg/kg).

κ).*Black currants.*—After treatment with endosulfan (I), black currants, as a rule, show relatively high residue levels (∿ 1 mg/kg) (*Hoechst AG* 1974). This finding was confirmed by Petrova *et al.* (1978). They found rather differing residue values which, however, did not exceed 2 mg/kg.

2. **Endosulfan residues in tea and tea infusions.**—The residue pattern of endosulfan on tea leaves was reported by *Hoechst AG* (1971). Tea shrubs, at two different altitudes of *PASI Tea Research Station* in Cinchona (India), were treated with Thiodan 35 EC at 1.25 or 2.50 L/ha, respectively. Table XXXI shows the average residues figures found. The residues, during preparation of the tea infusion, pass to a minor extent into the infusion. The maximum extraction rate was 30% of the residue in total contained in black tea. In preparing soluble tea powder, *viz.*, instant tea, only 5% of the total residues contained in black tea passed into the soluble tea powder [*cp.*, the report "Determination of endosulfan (α- and β-isomer) and endosulfan sulfate in tea infusions prepared from instant tea", *Hoechst AG* 1973].

3. **Endosulfan residues in green and roasted coffee, further in coffee infusions.**—There are a few unpublished papers of *Hoechst AG* (1973 and 1974) dealing with the residue behavior of endosulfan in coffee. In Brazil, Guatemala, and Cameroon, coffee shrubs (*e.g.*, var. Mundo Novo, 2.5 to 3.0 m high) were treated with Thiodan 35 EC emulsifiable. A survey of the residue results in Brazilian coffee is given in Table XXXII.

Tests with Thiodan 35 EC having an addition of Shell spray oil yielded higher amounts of residues, especially at higher application rates of 4 L/ha (maximum total residues 4.2 mg/kg 14 days after the last treatment).

Table XXXI. *Residue behavior of endosulfan in tea (India)* (*Hoechst AG* 1971).

1. Green sun-cured tea, 1.25 L/ha (2.50 L/ha = values in brackets)
   a) Altitude 1,500 m

Days after treatment	Endosulfan (mg/kg)[a]			
	αI	βI	-sulfate IV	Total
1	8 (11)	12 (16)	4 (6)	24 (33)
7	2 (1)	9 (3)	8 (6)	19 (10)
15	n.d. (0.6)	0.4 (2)	3 (2)	3.4 (4.4)
b) Altitude 900 m				
1	1.2 (4)	1.7 (8)	0.3 (2)	2.3 (14)
7	n.d. (0.1)	0.5 (1)	2 (3)	2.5 (4)
15	n.d. (n.d.)	n.d. (n.d.)	0.6 (0.5)	0.6 (0.5)

2. Fermented black tea, 1.25 L/ha (2.50 L/ha = values in brackets)
   a) Altitude 1,500 m

1	7 (21)	12 (34)	5 (8)	24 (63)
7	0.4 (3)	3 (4)	5 (5)	8.4 (12)
15	0.1 (0.3)	0.3 (1)	3 (4)	3.4 (5.3)
b) Altitude 900 m				
1	2 (5)	4 (9)	2 (3)	8 (17)
7	1 (3)	5 (13)	4 (7)	10 (23)
15	n.d. (1)	0.2 (3)	1 (4)	1.2 (7)

[a] n.d. = not detectable.

The degradation rate of the residues on green coffee after a treatment 4 times with 2 L Thiodan 35 EC is seen in Figure 16.

Further residue figures from field applications at the rate of 2 L/ha Thiodan 35 EC are at hand for the year 1974 from Brazil, Guatemala, and Cameroon. Out of a total of 19 samples (green coffee-beans) only a single one showed residues of 0.2 mg/kg (total residues) and two others contained residues between 0.05 and 0.7 mg/kg. All the rest of the samples were virtually free from endosulfan (I) residues (< 0.01 mg/kg). Waiting periods between the last treatment and harvesting were an average of 30 to 140 days, and the number of applications was one to two.

Ribas *et al.* (1974) also have found no residues after 100 days following a single treatment with endosulfan.

In roasting crude coffee containing endosulfan (I) residues, these latter are eliminated. No endosulfan ether (III) evolves in this connection (*Hoechst AG* 1974). Ribas *et al.* (1977) reported also on the influence exerted by roasting on the endosulfan residues, and they have come to the same results.

Table XXXII. *Residue behavior of endosulfan (Thiodan® 35 EC) in coffee (Brazil, Sao Paulo) (Hoechst AG, unpublished).[a]*

1. Altitude 1,000 m

Days after last treatment	Endosulfan (mg/kg)			
	αI	βI	Sulfate (IV)	Total
14	n.d.	0.02	0.04	0.06
	(0.09)	(0.07)	(0.07)	(0.23)
31	0.07	0.06	0.05	0.18
	(0.12)	(0.04)	(0.03)	(0.19)

2. Altitude 700 m

Days after last treatment	αI	βI	Sulfate (IV)	Total
14	n.d.	n.d.	0.02	0.02
	(n.d.)	(n.d.)	(0.07)	(0.07)
31	0.02	n.d.	0.07	0.09
	(n.d.)	(n.d.)	(0.01)	(0.01)

[a] Green air-cured coffee, 2 L/ha (4L/ha = values in brackets), two treatments in 21-day intervals (May-June).

[b] n.d. = not detectable.

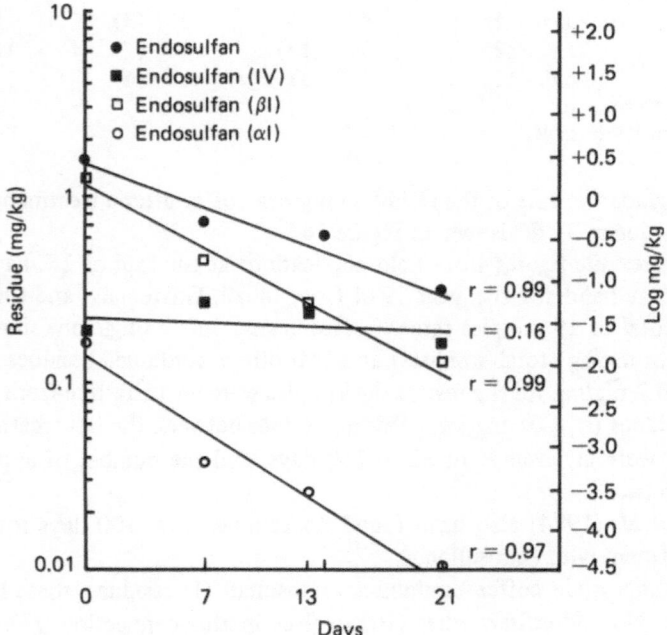

Fig. 16. Degradation rate of endosulfan on green coffee-beans (Brazil) (*Hoechst AG*, unpublished).

4. Endosulfan residues in tobacco.—Guthrie and Bowery (1967) presented a survey on residues from plant pesticides, including endosulfan, in tobacco up to 1967.

In the years prior to 1969, endosulfan was used in tobacco crops but in a minor degree. Parallel to the decreased applications of DDT and dieldrin, the endosulfan applications increased. Endosulfan residues in tobacco samples from auction sales and storehouses are, therefore, encountered in major numbers only from 1969 and later. Gibson *et al.* (1974) have given a good survey on this subject.

In the tobacco samples from the tobacco pool in Kentucky (U.S.A.), no endosulfan residues were found in the years 1963 to 1968. In 1969 for the first time residues of 0.8 to 0.9 mg/kg were detected, which increased to 1.4 to 3.4 mg/kg. The total residues of chlorine-containing plant pesticides on Burley tobacco in Kentucky consisted by 90% of endosulfan.

In the cigarette industry a current control for residues on imported tobacco is carried out. In the Federal Republic of Germany, a range for endosulfan (I) residues between "not detectable" (n.d.) and 9 mg/kg was found. In freshly cured "green" tobacco, up to 12 mg/kg were found (private communication of the scientific division in the *Verband der Zigarettenindustrie* 1975). A compilation of residue values in crude tobacco of different provenances is contained in Table XXXIV (*Werkhof GmbH* 1976). Further information on residue levels in tobacco samples from auction tobacco in the U.S.A. are given at the end of Table XXXIII and in Table XXXV.

A survey of the process of diminution of endosulfan after application under different vegetation and application conditions is demonstrated in numerous examples in Table XXXIII. The residue values given in Table XXXIII always represent the total residue consisting of endosulfan (αI and βI) and endosulfan sulfate (IV). Endosulfan sulfate (IV) always constitutes a relatively high proportion of the residue. A survey is shown in Table XXXV (Gibson *et al.* 1974).

Presumably endosulfan sulfate (IV) evolves preferably after the harvest when the leaves lie stacked until actual processing begins. The observations made by Archer (1973) during curing of lucerne allow this conclusion, the more so since tobacco, treated with endosulfan in the greenhouse and analyzed immediately after sampling, had only minor endosulfan sulfate (IV) concentrations (Gorbach and Schulze 1974). Further comments on the degradation of endosulfan (I) are found in the papers by Keil *et al.* (1972) and Cheng and Braun (1977).

A reduction of residues in harvested tobacco occurs especially by curing in an oven. Guthrie and Bowery (1962) have found a drop of the content in endosulfan residues by 83%, while Marais and van Wyk (1976) found a reduction by a maximum of 60%. The influence of freeze-drying of tobacco on the endosulfan residues was examined by Johnson *et al.* (1975). The residues were diminished by up to 42%. Although, *e.g.*, in Germany, the MRL proposal for endosulfan refers to tobacco products, no data are available on the residue level, *e.g.*, in cigarettes, and but one report on cigars (Domanksi and Guthrie 1974).

Table XXXIII. *Thiodan in tobacco (Hoechst AG, unpublished) (Niagara, unpublished).*

Location	Formulation	No. of applications	Time of last application			Application rate (kg/ha)	Analytical methods	Ref.	Residues (mg/kg) weeks after last application							
			Year	Month	Day				days			weeks				
									0	2-3	5	1	1.5	2	2.5	5
Quincey, Florida	4 Dust	9		5	27	0.67-0.84	GC	FMC	19.9	12.5	9.3	8.7				
Quincey, Florida	3 Dust	8		6	23	1.16	GC	FMC	51.6	42.0	24.6	23.4				
Clifton, Georgia	2 EC	4		6	22	1.12	GC	FMC	24.4	14.5	14.5	7.5				
North Carolina	2 EC	7		8	13	1.12	GC	FMC	76.7		54.1		20.0			
Madras, India	35 EC	1				0.525[a]	GC/ECD	H	56.0	14.4	0.6		0.4			
Madras, India	35 EC	2				0.525[a]	GC/ECD	H	17.8	6.6	3.5		2.6			
Madras, India	35 EC	3				0.525[a]	GC/ECD	H	37.4	15.2	14.8		11			
Madras, India	4 Dust	1				0.6[a]	GC/ECD	H	9.9	3.1	22.5		6.0			
Madras, India	4 Dust	2				0.6[a]	GC/ECD	H	53.5	37.2	7.5		7.7			
Madras, India	4 Dust	3				0.6[a]	GC/ECD	H	86.6	79.2	49.7		28			
Madras, India	35 WP	1				0.525[a]	GC/ECD	H	79.2	14.5	9.5		3.5			

Location	Form.	No.				Rate	Method	Compound							
Madras, India	35 WP	2				0.525[a]	GC/ECD	H	22.4	10.4	3.0				1.6
Madras, India	35 WP	3				0.525[a]	GC/ECD	H	46.2	14.8	11.4				6.5
Madras, India	35 EC	1				1.05[a]	GC/ECD	H	30.8	33.2	3.7				8.1
Madras, India	35 EC	2				1.05[a]	GC/ECD	H	22.7	13.2	3.2				3.3
Madras, India	35 EC	3				1.05[a]	GC/ECD	H	66.3	30.2	16.3				13.4
Princeton, Kentucky	2 EC	6	71	8	21	0.56	GC/MC	FMC[b]	25.0		3.8	3.3	2.0	2.2	
Princeton, Kentucky	2 EC	6	71	8	21		GC/MC	FMC[b]	8.5		7.0	7.6	11.7	12.2	
Princeton, Kentucky	2 EC	4	71	8	21	0.56	GC/MC	FMC[b]	107.0		15.0	15.3			11.0
Kentucky	2 EC	4	71	8	27	0.56	GC/MC	FMC[b]	15.0		11.9	10.8			8.6
Princeton, Kentucky	2 EC	5	71	8	21	0.56	GC/MC	FMC[b]	55.0		3.1	2.7	1.4	1.6	16.8
Princeton, Kentucky	2 EC	5	71	8	21	0.56	GC/MC	FMC[b]	14.0		13.5	15.4	14.8	15.2	22.2
Manisa, Orient	35 EC	2	73	7	30	1.05[a] each	GC/MC	H			12.1				
Denizli, Orient	35 EC	3	73	8	13	1.05[a] each	GC/MC	H							39.3
Ching mai	35 EC	6	71	2	15	2.08[a] each	GC/MC	H		9.5					
	35 EC	7	71	2	3	2.86[a] each	GC/MC	H							2.6

Table XXXIII (continued)

Location	Formulation	No. of applications	Time of last application Year	Month	Day	Application rate (kg/ha)	Analytical methods	Ref.	Residues (mg/kg) weeks after last application days 0	days 2-3	days 5	weeks 1	weeks 1.5
	35 EC	9	71	3	19	3.40[a] each	GC/MC	H		1.4			
	35 EC	10	71	2	10	1.99[a] each	GC/MC	H		8.6			
Quincy, Florida		4	65	6	22	0.83	GC/MC				28.2		
		2	65	6	22	0.9					18.4		
		2	65	6	22	1.1						33.	
1970 Auction Market U.S.A.							GC/MC	NC[b]	0.2-14.3 (3.1)				22
							GC/MC	NC[b]	0.3-12.5 (6.8)				10
							GC/MC	NC[b]	0.2- 3.3 (1.5)				6
							GC/MC	NC[b]	1.4-11.9 (4.3)				19
Georgia-Florida							GC/MC	NC[b]	0.2-11.1 (3.6)				24
Border							GC/MC	NC[b]	0.2-21.9 (3.9)				37
Eastern							GC/MC	NC[b]	0.3- 5.0 (1.5)				22
Middle							GC/MC	NC[b]	0.2- 4.5 (1.0)				8
Old							GC/MC	NC[b]	0.2- 2.7 (0.7)				2

[a] L/ha.
[b] NC = North Carolina State University; FMC = FMC Division, Niagara Chemical Corporation.

Table XXXIV. *Residue values of endosulfan in crude tobacco of different origins (WERKHOF GmbH 1976).*

Country	Crop	Minim. value[a] (mg/kg)	Max. value[a] (mg/kg)
China Flue-cured	1975	0.06	0.14
Indian Oriental	1974/75	0.10	0.20
Korea Burley	1975	0.01	0.43
Korea Flue-cured	1975	0.01	0.46
Philippine Flue-cured	1976	0.01	8.8
Taiwan Flue-cured	1976	n.d.	n.d.
Thailand Flue-cured	1976	0.02	3.1
Bulgarian Flue-cured	1975	0.01	0.01
Greek Burley	1975	0.01	0.60
Italian Burley	1975	0.02	0.56
UdSSR	1975	0.02	0.09
Brazilian Flue-cured	1976	n.d.	n.d.
Brazilian Bahia	1975/76	n.d.	n.d.
Colombian	1975/76	0.07	0.71
Guatemala Burley	1975	0.04	0.17
Panama Burley	1975	n.d.	n.d.
Chile Burley	1975	0.05	0.42

[a] n.d. = not detectable.

Guthrie and Bowery indicated for the first time in 1962 what amounts of residue of the ready-made cigarette pass during smoking into the main stream of smoke that is inhaled. They could demonstrate that as to endosulfan, such amount constitutes only 3% of the residue in tobacco. *Niagara Chemical Division of FMC* have come to a similar result in a study in 1971. Passage of up to 16%, however, was reported by Hengy and Thirion (1971). It was found by *Hoechst AG* (1975) that in smoking cigarettes with a deposit of 500 mg/kg of endosulfan, the main smoke-stream contains 11% endosulfan. Mass-spectrometric analysis of the smoke-condensate revealed that apart from endosulfan or endosulfan sulfate, respectively, no other chlorine-containing substances occurred ($< 10\%$ referred to the total chlorine content found in the condensate). The ^{14}C balance of the parent compound labeled in the methylene bridge confirmed the results.

The German cigarette industry, in view of the importance of paragraphs 1 and 2 of § 3 of *Höchstmengenverordnung Pflanzenbehandlungsmittel* (Ordinance on MRLs for agents for plant treatment) dated June 13, 1978 has proposed for endosulfan an MRL of 20 mg/kg in the finished tobacco product, which, mean-while, has been accepted by *Bundesgesundheitsamt* (*BGA*) (Federal Health Office).

5. **Endosulfan residues in the daily food.**—For some years endosulfan has been used in an appreciable amount in the U.S.A. and it was hence included in

Table XXXV. *Endosulfan components on Burley tobacco sampled from auction markets in Kentucky (Gibson et al. 1974).*

Crop year and insecticide	mg/kg and % of total on leaves from different portions of plant[a]				
	Top	Middle	Bottom	Average	Composite[b]
1970					
Endosulfan (αI)	0.65	0.80	0.86	0.77 (18.9)	0.78 (17.4)
Endosulfan (βI)	1.1	1.1	1.2	1.1 (28.5)	1.3 (30.8)
E. Sulfate (IV)	1.8	2.3	2.3	2.1 (52.6)	2.3 (51.8)
Total	3.6	4.2	4.3	4.1	4.4
1971					
Endosulfan (αI)	0.51	0.88	0.90	0.76 (18.5)	0.80 (16.1)
Endosulfan (βI)	2.5	1.0	1.5	1.7 (41.4)	2.2 (45.8)
E. Sulfate (IV)	1.4	1.4	2.1	1.6 (40.1)	1.8 (38.1)
Total	4.4	3.3	4.6	4.1	4.9
1972					
Endosulfan (αI)	0.29	0.31	0.66	0.42 (8.3)	0.25 (7.1)
Endosulfan (βI)	0.79	1.1	2.6	1.5 (16.1)	0.81 (23.0)
E. Sulfate (IV)	2.6	3.5	3.1	3.1 (75.6)	2.4 (69.9)
Total	3.6	5.0	6.4	5.0	3.5

[a] Average of all samples analyzed for indicated year.
[b] Sample consisting of equal quantities of top, middle, and bottom leaves.

the Pesticides Monitoring Program of the *U.S. Food and Drug Administration* covering several active substances.

Corneliussen (1969, 1970, and 1972), and later Manske and Corneliussen (1974), respectively Manske and Johnson (1975), Johnson and Manske (1976), and Manske and Johnson (1977) published their analyzed results from residues found in ready-to-eat foods. The foodstuffs were bought in, to begin with, 30, later on 35 markets in 24, respectively, later in 32 different cities with 50,000 to 1,000,000 inhabitants. A review of the results concerning endosulfan is found in Table XXXVI. The endosulfan (I) residues in food were predominantly due to residues deposited on fruits and vegetables.

In the years 1974 to 1975, in 7 of 240 samples of leaf-vegetables, 0.004 to 0.022 mg/kg of endosulfan residues were found, and in fruits there were traced up to 0.006 mg/kg of said residues.

It must be stressed that in fat and milk, in no case were residues detected. The average uptake of residues, therefore, amounted to 0.001 mg/person/day (Duggan and Corneliussen 1972).

In corresponding investigations in Canada (Smith 1971, Smith *et al.* 1972 and 1973), residues were found only in leafy-vegetables. Thus the average endosulfan residues in the years 1969, 1970, and 1971 were 0.006 mg/kg, 0.002, and 0.008 mg/kg, respectively. Investigations covering all regions of Winnipeg and Toronto showed a similar result (0.004 and 0.001 mg/kg on the average) (Smith *et al.* 1975).

Within the scope of the Canadian National Monitoring Program for Pesticides Residues in foodstuffs in the years 1972 and 1973, endosulfan residues were found in 8% of the apple samples (mean value 0.014 mg/kg) and in 4% of all bean samples (mean value 0.08 mg/kg) (Bluman 1973).

Table XXXVI. *Endosulfan residues in ready-to-serve foods (after Corneliussen 1969, 1970, and 1972; Manske and Johnson 1975; Johnson and Manske, 1976; Manske and Johnson 1977).*

Period	No. of ready-to-serve foods containing endosulfan residues, as compared to total no.	Residue range (mg/kg)
1967-1968	3 of 360	0.008-0.13
1968-1969	19 of 360	0.003-0.33
1969-1970	19 of 360	0.001-0.063
1970-1971	24 of 360	0.001-0.063
1971-1972	20 of 420	0.003-0.010
1972-1973	29 of 360	0.002-0.44
1973-1974	27 of 360	0.003-0.012

f) The influence on endosulfan residues by the processing
of food and stimulants

A summary of what is known of the influence of food processing on plant pesticide residues (including an example for endosulfan) until about 1968 was presented by Liska and Stadelman (1969) in Residue Reviews.

1. **Vegetable products.**—The decrease of endosulfan residues during processing of soybeans was the subject of a study by the *Niagara Chemical Division of FMC* (1969). Endosulfan (I) and endosulfan sulfate (IV) were distributed in approximately equal proportions between the crude oil and soybean meal. During refining of the soybean oil, the residues diminished by again 50% for endosulfan (I) and 20% for endosulfan sulfate (IV).

Results of a similar study concerning cottonseed are given in Table XXXVII.

In the processing of sugarbeets no enrichment of endosulfan (I) nor of endosulfan sulfate (IV) could be stated in the resulting sugarbeet draff. It may be remarked that already the initial residues in the sugarbeets after conventional application (3 applications of 1.1 kg/ha a.i. Thiodan EC by aircraft) are at the limit of detection (*Niagara Chem. Div., FMC-Corp.* 1971).

In processing grapes (with a residue of 0.25 mg/kg of endosulfan) to grape juice 0.58 mg/kg remained in the dried draff. Endosulfan sulfate was not detected in any of the samples in this study (*Niagara Chemical Division, FMC-Corp.* 1972).

Elkins *et al.* (1972) reported on the residue decrease of plant pesticides in the manufacture of heat-sterilized preserves of spinach and apricots. They also observed the further residue reduction during storage of the preserves. Table XXXVIII shows the results obtained for endosulfan.

The influence of the pickling brine on plant pesticide residues (including endosulfan) in the preservation of cucumbers was studied by Fehringer (1978). Endosulfan was completely degraded in said procedure.

Table XXXVII. *Endosulfan residues in cotton and residue distribution after processing (after NIAGARA CHEM. DIV. FMC-CORP. 1969).*

Product	Residue (mg/kg)	
	Test 1[a]	Test 2[b]
Seed	0.85	0.19
Wool	2.8	0.26
Wool dust	14	0.33
Husks	0.9	0.01
Crude oil	0.26	0.16
Refined oil	0.16	n.n.
Oil-cake	0.06	0.02

[a] Five applications of 3.3 kg/ha a.i. (Thiodan EC), sampling two days after last application.
[b] As in test 1, yet only 1.1 kg/ha a.i.

Table XXXVIII. *Residue reduction of endosulfan in spinach and apricots after preservation and storing (after Elkins et al. 1972).*

| Crop | Initial concentration (mg/kg) | Processed | Residue reduction (%) after 1 year storage | |
			At normal ambient temperatures	At 100°F (37.8°C)
Spinach	1.84	19	19	85
Apricots	1.79	13	22	85

Endosulfan residues (∿ 40 mg/kg) in water may be completely eliminated with 5 mg of permanganate/L (Bauer 1972). Blowing-out with ionized and non-ionized air led to an approximately 50% reduction of residues.

2. **Foodstuffs of animal origin.**—Endosulfan (I), even in high oral doses, does not pass into the meat, the fat, or the milk. Depending on the administered dose, however, the metabolite endosulfan sulfate (IV) is formed in the fatty tissue and in the milk (Gorbach et al. 1968, McCaskey and Liska 1967).

In order to study the effect of processing on the residues in the milk, McCaskey and Liska (1967) raised the endosulfan dose up to 2 g daily over 11 days and thus achieved a residue of 0.6 mg/kg of endosulfan sulfate (IV) (referred to crude milk). With processing, the residue decreased according to the indications given in Table XXXIX.

Li et al. (1970) reported on a similar trial. Endosulfan was fed in comparison with DDT, heptachlor, lindane, dicofol, and toxaphene to milchers. Only in traces could endosulfan be recovered in the milk and dairy products prepared therefrom, such as, e.g., cheese. As a matter of fact, endosulfan sulfate (IV) residues were not searched for.

Summary

A survey is presented of the behavior of endosulfan in the environment and of the occurrence of residues in food and forage crops.

The influencing factors are treated individually, and particularly the behavior in the soil and in surface waters is described. The residue behavior in foodstuffs is discussed, separately with regard to foodstuffs and forage of plant origin and stimulants such as coffee, tea, and tobacco.

It was demonstrated that the amounts of residues detected in samples from supervised trials do not allow any conclusions as to the virtually negligible amounts of endosulfan really taken up with the daily food.

A special paragraph is dedicated to the influence of processing operations on the residues. The reduction of residues is partly considerable in this connection.

Table XXXIX. *Reduction of endosulfan sulfate (IV) residues during processing of crude milk (after McCaskey and Liska 1967).*

Milk	Residue reduction (%) (referred to amount of residue in crude milk)
Crude	—
Condensed	—
Spray-dried	42
Evaporated	42
Kiln-dried	70

VII. Effect of endosulfan and its metabolites on arthropods

By W. Knauf

a) Introduction

Endosulfan is an insecticide/acaricide with primarily contact and stomach poison effect. No substantial transport within the vascular system of plants nor any so-called systemic action resulting therefrom has been observed so far.

A series of publications gives evidence of the remarkable potency of endosulfan against numerous insects and some species of mites in pest control. The results presented hereafter were mainly obtained from laboratory tests which allow quantification of the action. They are, therefore, subject to the usual reservations as to comparisons between laboratory and field trials; the examples given in Section IX, however, with regard to field application of endosulfan constitute a confirmation in all cases of said laboratory findings.

As will be shown, endosulfan covers a relatively broad spectrum of harmful arthropods, which emphasizes its character as a versatile plant protection agent.

b) Effect of endosulfan (I)

1. Contact effect.—In laboratory tests the topical application is used in most cases for investigating the contact poison effect of a substance, carried out in such a way that simultaneous oral intake of the active substance is excluded. To this effect additional so-called dish tests are conducted in which the inner surface of a Petri dish, or a substrate in such a dish, is treated with the active substance or its formulation (if need be, pre-dissolved in a solvent) in various concentrations and in conventional rates of application. Subsequently the test animals are placed onto the treated substrate and the Petri dish is closed. The oral intake, in these cases, is not always clearly to be separated from the uptake via the insect cuticula; similarly an action over the vapor (or gas) phase may play a role, unless the air above is constantly removed by suction.

Lindquist and Dahm (1957) investigated the insecticidal action of endosulfan (αI), of endosulfan (βI), and of purified endosulfan in topical application in comparison to DDT in houseflies (*Musca domestica* L.). By means of a pipette, one μL of an acetonic solution was applied onto the ventral thoracic region of female flies 3 to 5 days old. Then the treated insects were kept at 80° to 82° F (= 26.7° to 27.8°C) and 40 to 45% relative air humidity in cylindrical carton cages provided with sieve-like lids, and fed by means of cotton wadding soaked with sugar water and put on the lids.

The mortality was determined in each of 4 concentrations and in 3 to 4 repetitions with 20 to 50 flies each and evaluated in a dosage-mortality diagram (probit method) (Fig. 17). The authors determined with this method the LD_{50} (24 hr) of endosulfan (I) at 0.15 μg/fly and 6.7 μg/g fly, respectively.

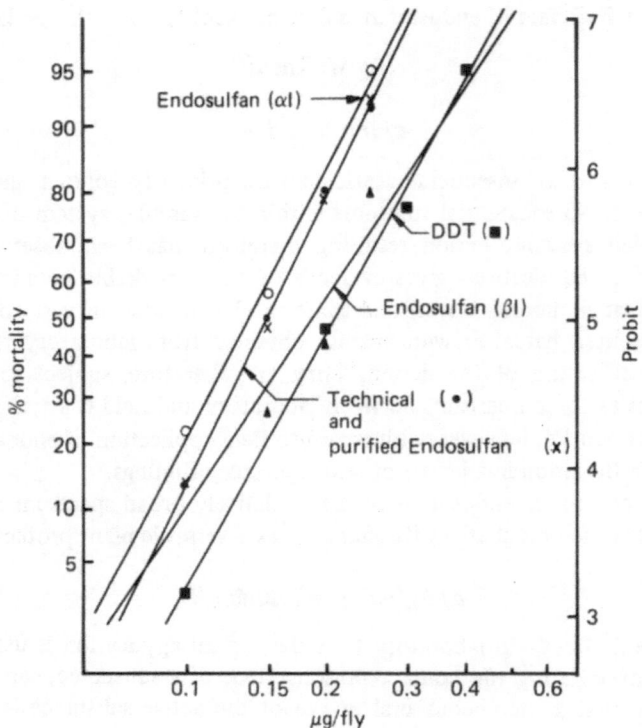

Fig. 17. Dosage-mortality curves comparing the action of technical Thiodan®, purified endosulfan isomers α and β, and p,p'-DDT to female houseflies (after Lindquist and Dahm 1957).

Almost the same values were obtained by Yun-Pei Sun (1972), who was able to ascertain the LD_{50} for endosulfan (I) for *Musca domestica* L. as 6.2 µg/g fly.

Keiser and Tohikawa (1970) checked the toxicity of endosulfan (I) in other species of *Diptera*. They found as LC_{50} (better termed as LD_{50}) in the Mediterranean fruit fly (*Ceratitis capitata* Wied.) after 24 hr 0.25 µg/g fly; for the Oriental fruit fly (*Dacus dorsalis* Hend.) 1.23 µg/fly and for the melon fly (*Dacus cucurbitae* Coq.) 0.075 µg/fly.

Here, too, one µL of a solution of endosulfan (I) in acetone was brought onto the thorax (mesonotum) by means of a microliter syringe.

Under laboratory conditions Lal (1971) tested flies of the species *Phytomyza atricornis* Meig. with the Petri dish method. They were placed into Petri dishes that had been treated by means of a Potter tower with one ml each of an aqueous emulsion of Thiodan 35 EC in varying concentrations. The dishes set with flies after drying were then placed under observation for 48 hr at 27° ± 1°C and 75% relative air humidity (Fig. 18).

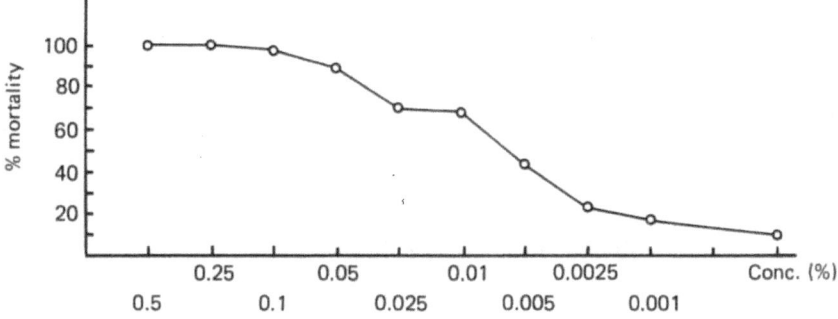

Fig. 18. Effect of different concentrations of Thiodan® 35 EC on the adult mortality of *Phytomyza atricornis* Meiq. in two days (Lal 1971).

Tests with pupae placed in Petri dishes that had been treated in the same way showed within a period of 7 days the mortality rates illustrated in Fig. 19. Any action via the gas phase could not be excluded.

Roemer (1957) has carried out investigations concerning the contact action of endosulfan (I), in comparison to lindane, in Colorado beetles (*Leptinotarsa decemlineata* Say). For this purpose the oral orifices of the beetles were closed with an inert adhesive and endosulfan (I) was applied onto the tarsi. Roemer operated with doses of 0.5 to 0.008 μg/animal, the quantity of active substance used in each case (dissolved in acetone) being applied by distribution upon all tarsi. The test period was 14 days, the temperature 23°C (Table XL).

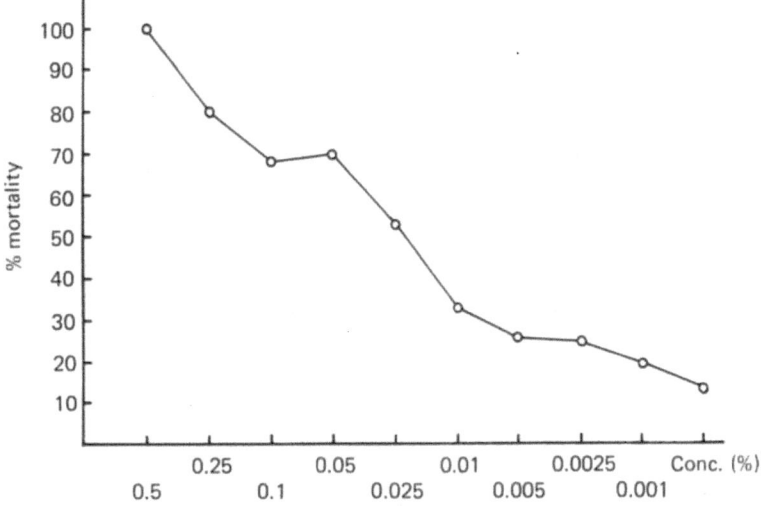

Fig. 19. Pupal mortality of *Phytomyza atricornis* Meiq. as effected by different concentrations of Thiodan® 35 EC in 7 days (Lal 1971).

Table XL. *Effect of endosulfan (I) and lindane on Leptinotarsa decemlineata (Colorado beetle) in comparison* (after Roemer 1957).

Compound		LD_{50}	LD_{100}
Endosulfan (I)	μg/animal	0.17 (0.12-0.24)	0.50 (0.25-1.05)
	mg/kg animal	1.42 (0.99-1.99)	4.17 (2.02-8.75)
Lindane	μg/animal	0.14 (0.11-0.17)	0.30 (0.19-0.49)
	mg/kg animal	1.17 (0.92-1.42)	2.80 (1.58-4.08)

In laboratory tests, the contact action of endosulfan (I) could be studied in grain weevils (*Sitophilus granarius*) on prepared paper sheets (Japan paper). For this test, 20 insects each per concentration were placed in Petri dishes laid out with the said paper, and they were then sprayed by means of a spraying apparatus with gradually varying concentrations of an aqueous emulsion of Thiodan 35 EC (application rate corresponding to 600 l/ha). The covered Petri dishes were stored for 48 hr at 21°C and the mortality of the beetles was then determined (Fig. 20).

In like manner 20 larvae each of yellow mealworm (*Tenebrio molitor* L.) were treated in covered Petri dishes and stored for 48 hr (*Hoechst AG*, 1973). Any action via the gas phase could not be excluded either in these trials.

Olinger and Kerr (1969) ascertained the contact poison effect of endosulfan (I) by topical application on the sternites of larvae of the Mexican bean beetle (*Epilachna varivestis* Muls.) in the fourth instar. They used three different solvents:

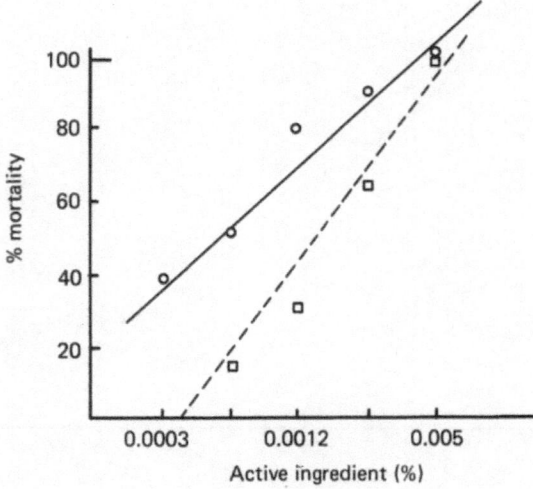

Fig. 20. Effect of endosulfan (I) on *Sitophilus granarius* L. (——) and *Tenebrio molitor* L. (– – –) in dish test (*Hoechst AG* 1973).

a) Acetone

b) DMSO (dimethyl-sulfoxide)

c) a mixture of both solvents in the ratio 1:1

The 1% solutions were applied with a micro-applicator in doses of one μl/animal; this corresponds to a dose of 0.01 mg/animal (3 repetitions, 60 animals/concentration). Subsequently the mortality of the larvae was determined in intervals of time (Fig. 21).

These tests demonstrate that suitable solvents (such as DMSO) may increase the rate of action of endosulfan (I) (presumably by improvement of the penetrating velocity into the cuticula).

Suber *et al.* (1971) compared the action of different insecticides following topical application on the cowpea curculio (*Chalcodermus aeneus* Boh.). They applied the active substances, dissolved in acetone, with a micro-applicator in equal doses of 2 μL in different concentrations on the ventral seam of the thorax and the abdomen.

After the treatment, the beetles were placed in Petri dishes, fed with bean leaves, whereupon the mortality was determined after 96 hr (50 insects/dose). The mortality values were calculated after the method of Litchfield and Wilcoxon (1949) (Fig. 22).

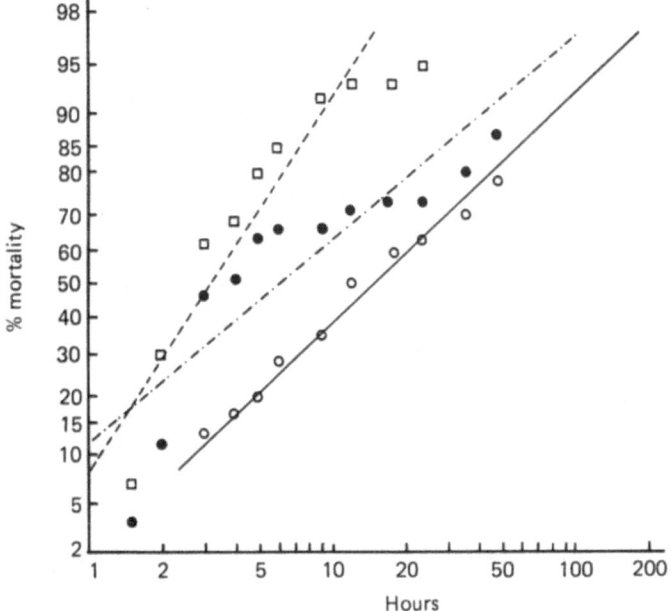

Fig. 21. Dosage-mortality regression lines for 1% endosulfan in acetone (——), DMSO (– – –), and a 1:1 mixture of the solvents (– · – ·) against 4th-instar larvae of *Epilachna varivestis* Mals. (Olinger and Kerr 1969).

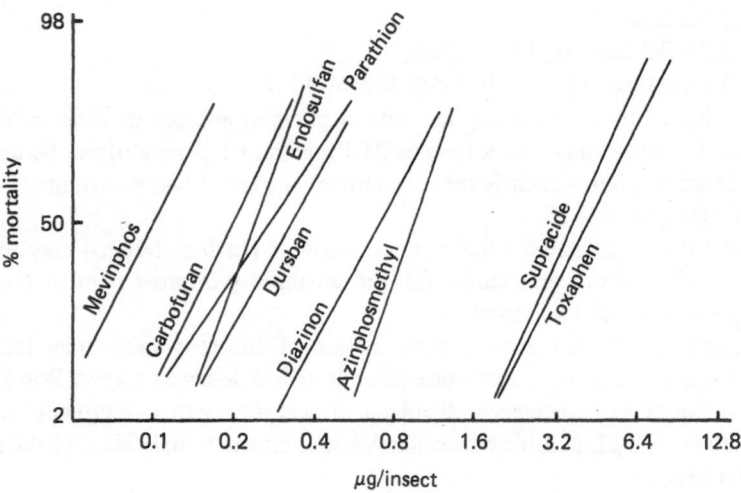

Fig. 22. Toxicities of various insecticides (topical application) to *Chalcodermus aeneus* Boh. (Suber *et al.* 1971).

Roemer (1957) investigated the contact poison effect of endosulfan (I) in the American cockroach (*Periplaneta americana* L.) by direct application of an acetone solution of the active substance upon the tarsi. The oral orifice of the test insects was closed with an adhesive. Here, too, as in the experiments with *Leptinotarsa decemlineata* Say doses were used ranging between 4 and 0.3 μg for male and between 11 and 1 μg for female insects. The mortality was determined after 14 days (temperature: 22° to 24°C) (Fig. 23). There resulted a higher susceptibility of male insects to endosulfan (I).

Further studies were conducted with the German cockroach (*Phyllodromia germanica* L.). Firstly, endosulfan technical was dissolved in acetone, and 2 ml each of the solution were evenly applied with a pipette to the inner surface of the bottom and lid of a Petri dish. The dish halves were dried for 1 hr (T = ~ 22°C). In this process evidently minor amounts of active substance evaporated, which means that the indications concerning surface concentration have to be considered as approximative.

After introducing 10 larvae (0.9 to 1.1 cm body length) of the previously CO_2-narcotized cockroaches, the closed dishes were placed at temperatures between 20° to 22°C, and after 3 days the mortality was determined (3 × 3 repetitions of 10 insects each/concentration (*cp.*, Fig. 24) (*Hoechst AG* 1973).

It has been found that a surface coating of 0.0243 mg/100 cm² is required for a 50% kill of the test animals. In this experiment, the contact action on the tarsi and the total cuticula (sublimation) as well as the gas phase was effective in the closed Petri dishes. On the strength of the experimental set-up the action should be produced also by the precipitation of the active substance on the cuticula (Schulze 1965). Additionally the active substance can act by direct contact over the tarsi.

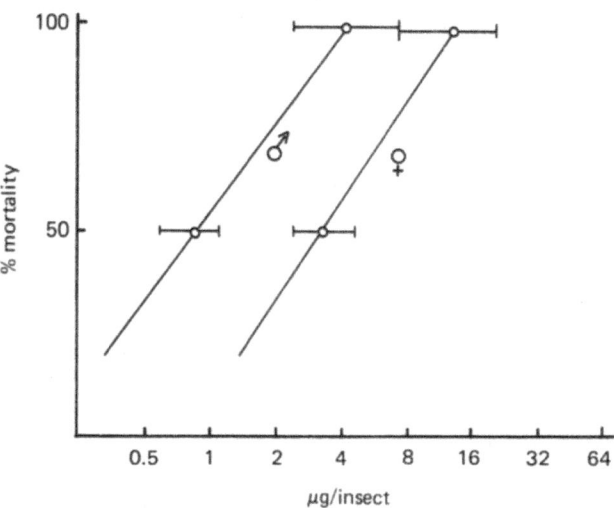

Fig. 23. Mortality of *P. americana* L. after tarsal application of endosulfan
isomers (αI) + (βI) (14 days) (after Roemer 1957).

Endosulfan (I) possesses a good action against hoppers. For tests with brown
planthoppers (*Nilaparvata lugens* Staol), imagines from laboratory breedings
were set into plastic Petri dishes the bottoms of which were lined with rice
paper. This paper was sprayed with different concentrations of an aqueous
Thiodan emulsion (applied quantity of emulsion/dish $\cong$ 600 L/ha). Subsequently
every 20 hoppers/concentration and dish were counted (4 repetitions) into the

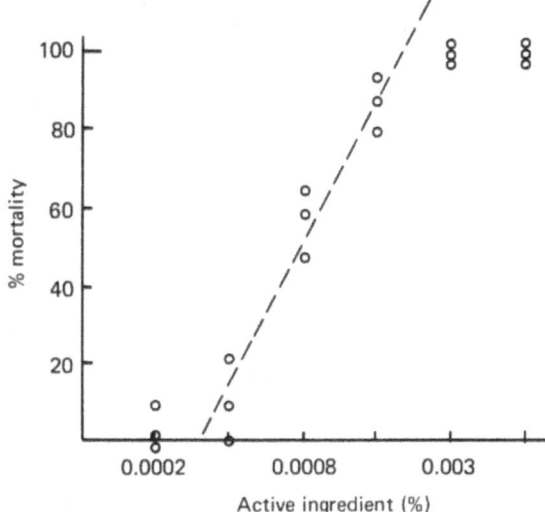

Fig. 24. Effect of endosulfan (I) on *Phyllodromia germanica* L. in the dish test
(*Hoechst AG* 1973).

dishes and these were placed at 21°C for 48 hr in the laboratory (Fig. 25). Any action over the gas phase or sublimation could not be excluded in this case either.

In experiments in glass dishes ($\emptyset$10 cm, height 6 cm), a good effect of endosulfan (I) was likewise obtained against adults of the European corn borer (*Pyrausta nubilalis* Huebn. images). After spraying of the glass dishes with an aqueous emulsion of a 30% Thiodan EC in different concentrations (application rate corresponding to 600 L/ha) the insects were set onto the dried coating and the dishes were covered with a permeable wire screen lid in order to reduce any gas phase effect. The mortality of the moths defined above determined after 24 hr is given in Table XLI (*Hoechst AG* 1973).

In tests with the aphid species *Doralis fabae, Myzus persicae* Sulz., and *Aulacorthum solani*, affected *Vicia faba* (*Doralis fabae*), and *Capsicum annuum* (*Myzus persicae* and *Aulacorthum solani* Kaet.) were used.

The plants for this test were sprayed with aqueous dilutions of Thiodan EC in the concentrations 0.01 to 0.00005%, until the plants were in the initial dropwet stage, then they were placed in a greenhouse (mean temperature 20° to 22°C) for a day-period of 16 hr. Subsequently the mortality was evaluated after Abbott *et al.* (1969) (Fig. 26).

2. **Feed effect.**–The orally effective dose of endosulfan has been determined more closely for some insect species.

Roemer (1957) orally applied to American cockroaches (*Periplaneta americana* L.) endosulfan (I) and Thiodan 20 EC dissolved in olive oil by means of a

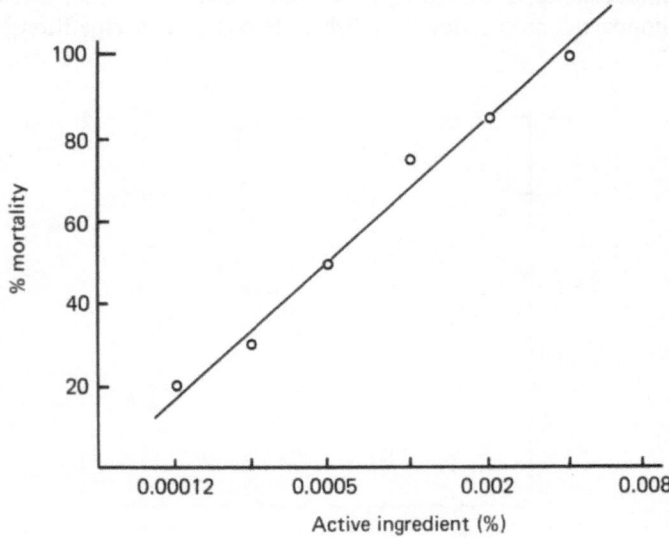

Fig. 25. The effect of endosulfan (I) on *Nilaparvata lugens* Staol. in the dish test (*Hoechst AG* 1973).

Table XLI. *Mortality of Pyrausta nubilalis Huebn. (European corn borer) on glass dishes treated with endosulfan (Hoechst AG 1973).*

% a.i.	% Mortality
0.005	100
0.0025	70
0.001	50
0.0005	20
0.00025	0

pharyngeal tube and observed the test insects for 20 days at 22° to 23°C. Twelve to 16 hr after the application, the first intoxication symptoms (hyperactivity, Schulze 1965) could be ascertained. The LD_{50} values (after 21 days) are given in Table XII.

It can be noted that the addition of emulsifiers has no influence on the effect of endosulfan (I) in male and female insects. It was, however, apparent that the

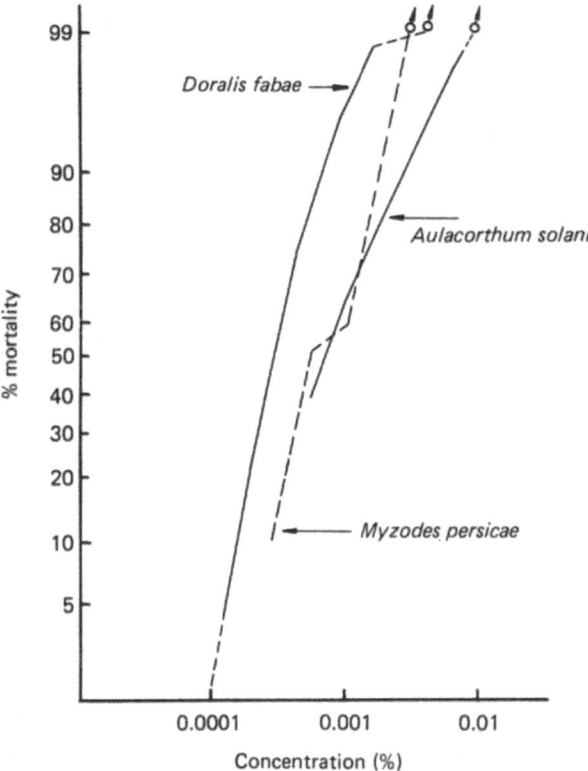

Fig. 26. Effect of Thiodan® 35 EC on aphids after three days (*Hoechst AG 1972*).

Table XLII. *Toxicity from endosulfan in Periplaneta americana* L. *after oral application* (after Roemer 1957).

Applied as	Sex	LD_{50} (21 days) (mg/kg a.i.)
Thiodan® 20 EC	♂	0.93 (0.87-1.16)
Endosulfan (I), purified	♂	0.88 (0.7-1.1)
Thiodan® 30 EC	♀	2.31 (1.64-3.23)
Endosulfan (I), purified	♀	2.31 (1.85-2.88)

females possess a higher tolerance to endosulfan (I), which was also observed by Schulze (1965) in the South American giant cockroach (*Blaberus trapezoides* Burm.).

Keller and Liang (1962) determined likewise in *Periplaneta americana* L. the oral toxicity of endosulfan (I) in solution in purified kerosene (deo-base). The authors introduced the cannula of a micro-applicator into the oral cavity of the insect and slowly injected 10 $\mu\ell$/insect, the solution being immediately swallowed into the gastrointestinal tract. After 24 hr the mortality was determined (sex of insects not indicated). As LD_{50} a value of 45.5 μg/g of cockroach was calculated.

Roemer (1957) used another application method with adults of the Colorado beetle (*Leptinotarsa decemlineata* Say imagines). The test beetles were offered endosulfan (I) sandwiched between 2 slices of potato leaves (sandwich method). Here, too, the observation period was 21 days and the test temperature 22°C (80 beetles, doses between 0.08 and 0.025 μg/beetle). The LD_{50} was found to be 0.35 (0.30 to 0.42) mg/kg.

Goesswald (1958) was able to determine the approximate oral toxicity in common black ants (*Lasius niger* L.) by indirect administration of 0.5 to 1% Thiodan 20 EC in honey water, making use of the social behavior of feeding each other within an integrated community (filter paper formicaries).

In order to separate as completely as possible the oral intoxication from the contact effect, Goesswald (1958) fed endosulfan in honey-water emulsion at first to the female worker ants who, on their part, conveyed their crop contents by mouth-to-mouth feeding to unfed ants. The honey-water emulsions were labelled with radioactive ^{32}P (as phosphate), and the uptake of the solution was observed in each single ant by means of a Geiger counter. The crop content of an ant was maximum 40 μg referred to the original emulsion.

The author ascertained that 10 μg/ant [corresponding to 0.01 μg of endosulfan (I)] does not cause any injury to the ants within 4 days, 12 to 15 μg [corresponding to 0.012 to 0.015 μg of endosulfan (I)] caused after 2 to 4 hr reversible injuries (coordination disorders, paralyses). A dose of 15 to 20 μg [corresponding to 0.015 to 0.02 μg of endosulfan (I)] or 20 μg, respectively, caused after 2 hr a "knock-down" effect and irreversible injuries.

Also in lepidopterous larvae with the red-backed cutworm (*Euxoa ochrogaster* Guen.) as example, trials for oral action of endosulfan have been carried out in the laboratory. McDonald (1972) fed pieces of lettuce that had been treated with endosulfan (I) dissolved in acetone to larvae of this species in the 5th instar. He used three different rates: 1.5, 0.25, and 0.025 µg/larva. The treated larvae remained for 72 hr at 24° ± 1°C and 50% relative atmospheric humidity in the observation vessels. The specimens that had not eaten the total quantity of leaf mass within 6 hr were exempted from evaluation; the animals that did not take up any food after the test were designated as "dead". The results (*cp.*, Table XLIII) reveal that the lethal dose for the larval stage L 5 lies between 2.5 and 0.25 µg/animal.

3. **Combined feed and contact effect.**–Laboratory tests with various lepidopterous species (*Plutella maculipennis* Curt., *Prodenia litura* Auct., *Pieris brassicae* L., and *Heliothis virescens* F.) showed 100% mortality after 48 hr at spray liquid concentrations between 0.01 and 0.05 % a.i. (*Hoechst AG* 1973). The larvae of the diamondback moth (*Plutella maculipennis* Curt. (L 4) stemmed from constant laboratory breeding on charlock (*Sinapis alba*), the African cotton leafworm (*Prodenia litura* Auct. (L 4)) from breeding on the bean *Vicia faba*, the cabbageworm (*Pieris brassicae* L.) had been raised on cabbage (*Brassica oleracea*), and the tobacco budworm (*Heliothis virescens* F. (L 4)) on an artificial nutrient medium.

The applications were carried out by means of a spray apparatus while using 0.1 to 0.00125 % a.i. in the spray broth. The larvae as well as the substrates were sprayed (*Plutella maculipennis* Curt. and *Pieris brassicae* L. / cabbage leaves, *Prodenia litura* Auct. and *Heliothis virescens* F. / cotton leaves). Subsequently, after drying, the larvae were placed on the substrates which were then put into wax beakers (contents 200 ml) that were subsequently covered with a screen lid so that a certain degree of air circulation was assured.

In these tests the effect from endosulfan is ascribed to a combined feed (stomach) and contact action, which simulates the natural conditions in the field (Fig. 27).

Founk and McLanahan (1970) treated Colorado beetle larvae (*Leptinotarsa decemlineata* Say) in a spray test with endosulfan in different aqueous concentrations by means of a "Potter tower applicator". The larvae were then kept on likewise-treated leaves in a climate closet. Figure 28 demonstrates the mortalities after 48 hr as compared to other preparations.

The effect of endosulfan (I) on bugs of the species *Oncopeltus fasciatus* Dall. (large milkweed bug) is shown in Figure 29. The adult insects used for this investigation originated from laboratory breeding on *Vicia faba* with additional feeding with sunflower seeds (*Hoechst AG* 1973). Ten bugs each on a nonabsorbent base were sprayed with an aqueous emulsion of endosulfan (I) in varying concentrations (quantity corresponding to 600 L spray liquid/ha). After the drying of the sprayed coating, the test animals were transferred into a beaker where they were fed with shoots of *Vicia faba*. To ensure a certain degree of air ex-

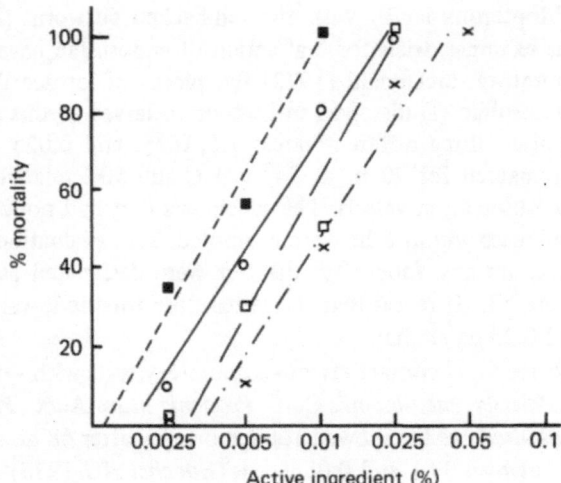

Fig. 27. Effect of endosulfan (I) on lepidopterous larvae (*Hoechst AG* 1973):
—— *P. litura* Auct., ——— *P. maculipennis* Curt., —·—·— *H. virescens* F.,
and - - - - *P. brassicae* L..

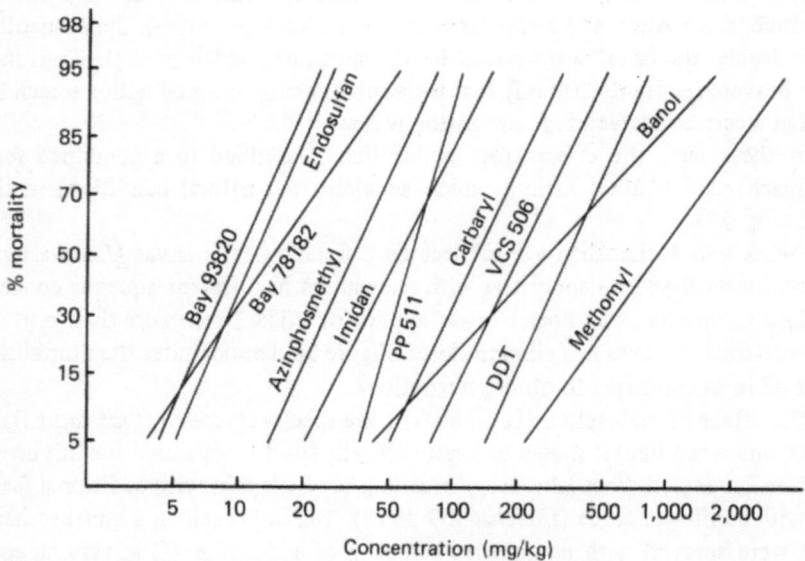

Fig. 28. Effect of endosulfan (I) as compared to other insecticides on *Leptiontarsa decemlineata* Say (after Founk and McClanahan 1970).

Table XLIII. *Effect of endosulfan (I) on Euxoa ochrohgaster* Guen. *after oral application on lettuce* (after McDonald 1972).

	Dose (µg/insect)	% Mortality after		
		24 hr	48 hr	72 hr
Endosulfan	2.5	100	100	100
	0.25	40	40	47
	0.025	0	0	0

change, the paraffin-coated cardboard cups were covered with coarse-meshed tissue. After 5 days the mortality was determined (two repetitions).

c) Comparative effect of the single isomers of endosulfan (I) and its metabolites

Barnes and Ware (1965) conducted a comparative study of the toxicity of endosulfan (αI) and (βI), endosulfan sulfate (IV), endosulfan diol (II), and endosulfan ether (III) in houseflies (*Musca domestica* L.). They found that endosulfan (αI) was insecticidally more potent (1.63x) than endosulfan (βI). The metabolite endosulfan sulfate (IV) corresponded in these trials in its activity against 3-day old *Musca domestica* L. approximately with that of endosulfan (βI). Endosulfan diol (II) and endosulfan ether (III) did not show any provable insecticidal property (Table XLIV).

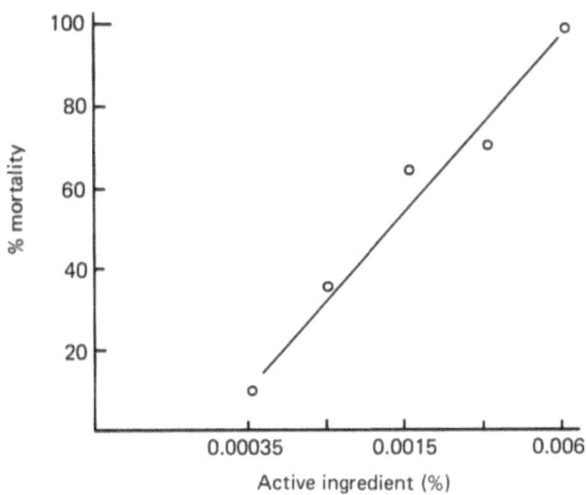

Fig. 29. Effect of endosulfan (I) on *Oncopeltus fasciatus* Dall. (*Hoechst AG* 1973).

Table XLIV. *Comparative effect of endosulfan and its metabolites on Musca domestica* L. *(houseflies)* (after Barnes and Ware 1965).

Compound	LD_{50} (μg/g body wt)
Endosulfan (αI)	5.5
Endosulfan (βI)	9.0
Endosulfan sulfate (IV)	9.5
Endosulfan diol (II)	> 500
Endosulfan ether (III)	> 500

Lindquist and Dahm (1957) determined the toxicity of endosulfan (αI) and endosulfan (βI) following topical application to *Musca domestica* L. (Table XLV).

Schulze (1965) observed the stages of intoxication with endosulfan (αI) and endosulfan (βI) in the grain weevil (*Sitophilus granarius* L) The test vessel was kept at a relative atmospheric humidity of 74% by means of a standardized saline solution. Intoxication was carried out by interaction in the "gas phase" by using a filter paper onto which 1.8 mg of the active substances dissolved in acetone were given dropwise and which after evaporation of the solvent was fixed, freely suspended, in the interior of the test vessel (Fig. 30).

It was concluded from this that endosulfan (αI) is essentially faster in its action than is endosulfan (βI) [this is explained by faster evaporation of endosulfan (αI)]. The same author states that to accomplish a defined stage of intoxication by endosulfan (αI), smaller amounts of the active substance should evidently be involved in the interior of the body (haemolymph) than in the case of endosulfan (βI). It must be assumed that endosulfan (αI) rather than endosulfan (βI) is faster in penetrating the integument into the body of *Sitophilus granarius* L.

In the cockroach *Phyllodromia germanica* L., the effect of endofulfan sulfate (IV), endosulfan diol (II), endosulfan ether (III), and endosulfan lactone (VI) were examined by comparison. Intoxication occurred in all cases in glass Petri dishes whose bottoms or lids, respectively, received two ml each of different pesticide solutions by means of a pipette. Ten larvae (approximately 8 mm in length) each/concentration were set into the dishes, which were then observed

Table XLV. *The effect of the two endosulfan isomers and of DDT on Musca domestica* L. (after Lindquist and Dahm 1957).[a]

Compound	LD_{50} (μg/fly)	μg/fly
Endosulfan (αI)	0.14	6.2
Endosulfan (βI)	0.19	8.5
DDT	0.21	9.4

[a] See also Figure 17.

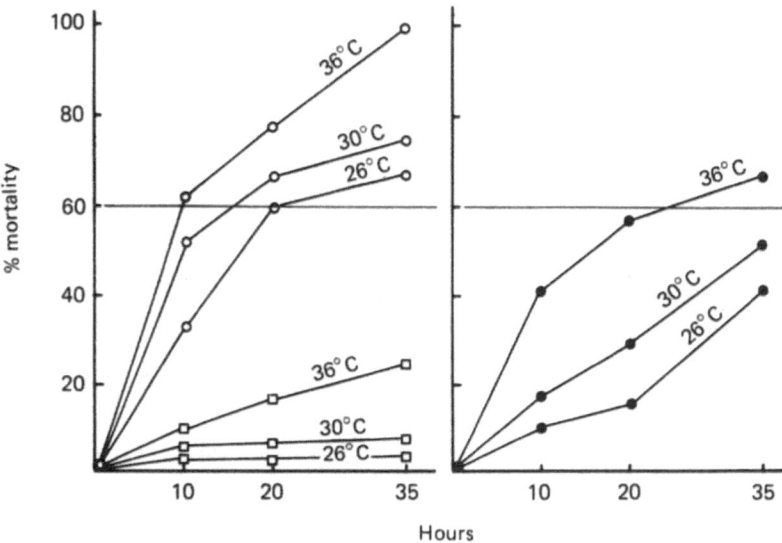

Fig. 30. Graphic demonstration of efficiency of Thiodan isomers at different temperatures in standardized relative atmospheric humidity of 74% (Schulze 1965): ○=endosulfan (αI), ● = endosulfan (βI), and □ = control.

at room temperature (21° to 22°C) for 3 days. The plates were closed during the experiment (*Hoechst AG* 1973).

As shown in Fig. 31, endosulfan (αI), rather than endosulfan (βI), proved to be the more potent isomer. Endosulfan sulfate (IV) is far less potent than endosulfan (βI).

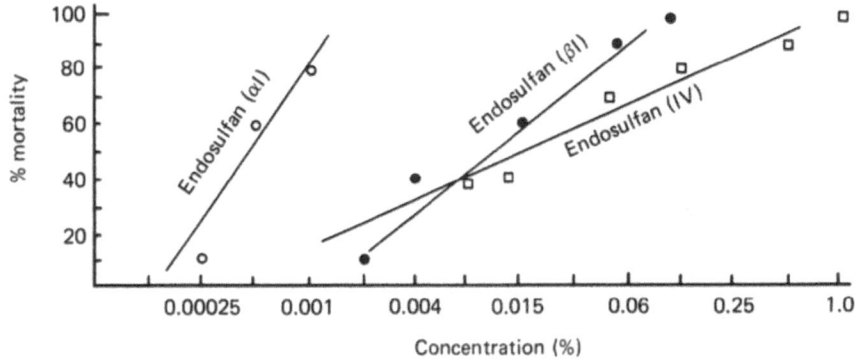

Fig. 31. Effect of endosulfan (αI and βI) and endosulfan sulfate (IV) in *Phyllodromia germanica* L. (*Hoechst AG* 1973).

The shallow gradient of the dose-mortality curves in the case of endosulfan sulfate (IV) suggests that this metabolite, therefore, reaches the site of activity later than endosulfan (βI). The metabolites endosulfan diol (II), endosulfan ether (III), and endosulfan lactone (VI) showed no mortality even in higher doses (10% acetone solution, 2 ml applied amount).

d) External intoxication symptoms

Klee (1960) distinguishes in endosulfan intoxication of the yellow-necked dry-wood termite (*Calotermes flavicollis* F.) and the grain weevil (*Sitophilus granarius* L.) 7 stages in intoxication which are consecutively passed through (Table XLVI).

Schulze (1965), however, distinguishes 6 different intoxication phases. These phases are observed in *Sitophilus granarius* F. as well as in *Tenebrio molitor* L., *Periplaneta americana* L., *Blaberus trapezoides* Burm., the pine sawfly (*Diprion pini* L. larvae), and in Gypsy moth larvae (*Lymantria dispar* L.).

Table XLVI. *Intoxication phases in insects after treatment with endosulfan* (Klee 1960).

Stage 1. Latent phase
At first the behavior of intoxicated insects does not differ from that of nonpoisoned insects. Movements are normal.

Stage 2. Excitation phase I
The insects become increasingly excited. Unrest leads to hasty movements. Restless to-and-fro running. Frequently wrong succession of legs.

Stage 3. Excitation phase II
Trembling uncertain movements of legs causing transitory lateral position. As if under raised inner pressure, the abdominal segments are pressed apart, followed by general defecation.

Stage 4. Noxious phase
This is characterized by irreversible supine position with still lively movements of body and extremities.

Stage 5. Anesthetic phase
General exhaustion. Insects are incapable of translocation.

Stage 6. Tremor phase
Arching of body, convulsive tremors of legs.

Stage 7. Adynamic phase
Almost complete immobility. To strong stimuli, there are only minor defense reactions followed by complete rigidity. Collapse of abdomen.

Since the intoxication symptoms in the above-mentioned test insects show a uniform pattern, it must be assumed that they are essentially similar in insects in general (Table XLVII).

e) Physiological responses of treated insects

Extensive studies to influence the physiological condition of insects after treatment with endosulfan were carried out by Goesswald (1958), Klee (1960), Goesswald (1962), Goesswald et al. (1963), Schulze (1965), and Schulze (1967). In addition to these, no papers of importance have appeared on this subject in recent times. There follows a review of the essential results of the authors referred to above.

1. **Oxygen consumption.**—A first hint to the increased oxygen consumption of yellow-necked dry-wood termites (*Calotermes flavicollis* F.) was given by Goesswald (1958). He found with the Warburg method that the animals, after a 3-hr contact with a blotting paper soaked with 0.6 ml of an aqueous 1% endosulfan emulsion 20 EC (= 0.12 mg a.i.) showed a more than three-fold increase of O_2-consumption as compared with controls. In order to know whether it is possible that by the increased motoric activity of the animals an indirect rise of the O_2-consumption may ensue, Goesswald investigated the respiration intensity of animals after a 6-hr contact poisoning with the contaminated filter paper. Here the animals were in a so-called "k.o. phase", *i.e.*, in supine position with only minor simultaneous movements of the extremities and the antennae. In spite of this a more than 2.5-fold increase of the oxygen consumption was found.

Klee (1960) ascertained in a Warburg vessel the O_2-consumption from the moment of beginning contact of *Calotermes flavicollis* F. with a layer of endosulfan on a glass surface. To this effect he rubbed a drop of a 20% Thiodan emulsion concentrate on the bottom of the Warburg vessel and then placed the test insects onto the coating.

In larvae of the bluebottle fly (*Calliphora erythrocephala* Meiq.) Klee could not observe any alteration of O_2 consumption during the test period, the test animals regularly consumed oxygen. In *Calotermes* he observed a two-fold increase of O_2-consumption following intoxication. By investigations in tethered and untethered larvae of said fly species, a comparison of the results revealed also an increased respiration through the skin. The author found that the O_2 consumption, after placing the animals on to the coating, started to rise immediately and slowly declined after rapid reaching of a maximum (three-fold of the controls) (Fig. 32).

Schulze (1965 and 1967), too, used the Warburg technique for ascertaining the O_2 consumption in contact-intoxicated *Sitophilus granarius* L., *Lymantria dispar* L. (larvae), and *Diprion pini* L. (larvae). He could confirm in *Sitophilus* the above described course of O_2 consumption. It follows therefrom that a

Table XLVII. *General intoxication symptoms* (Schulze 1965).

Intoxication stage	Sitophilus and Tenebrio	Periplaneta and Blaberus	Diprion	Lymantria
1 Latent phase	Normal behavior, beetles run about	Normal behavior (cleaning movements). Cockroaches mostly remain quiet in one place	Normal behavior, caterpillars creep or sit quietly	Normal behavior, caterpillars creep or sit quietly
2 Excitation phase I	Paralyzed tarsi on leg pair I, → increasing unrest → uncoordinated running	At first insensible to tactile stimuli → spontaneous motility, growing restlessness → abnormally panicky	Erection of fore and hind part → swelling of thoracic region → growing restlessness with hasty creeping movements	Growing restlessness with hasty creeping movements
3 Excitation phase II	Motoric hyperactivity → defecation, irreversible supine position with violent struggling of extremities	Stilted walking → motoric hyperactivity → irreversible supine position with violent struggling of extremities	Total body strongly contracted → motoric hyperactivity → discharge of body fluid from mouth and anus → irreversible lateral position with snake-like contractions	Caterpillars strongly contracted → motoric hyperactivity → fluid discharge from mouth and anus → irreversible lateral position with contractions
4 Tremor phase	Convulsive tremor of extremities	Convulsive tremor of extremities → oral organs in constant	Weak tremor of extremities → spontaneous body contractions	Weak tremor of extremities → spontaneous body contractions
5 Adynamic phase	Complete immobility → reactions of extremities and palps only on strong stimuli	Complete immobility → reactions of extremities, palps, and cerci only after strong stimuli	Complete immobility → distinct body contractions → dorsal vessel beats only irregularly	Complete immobility → distinct body contractions
6 Exitus	—	—	—	—

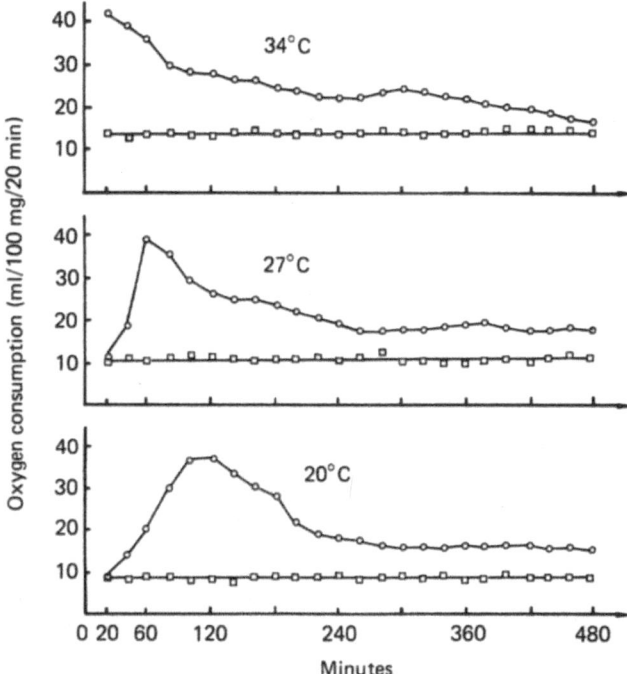

Fig. 32. O_2 consumption of *Calotermes flavicollis* F. after contact with endosulfan (Klee 1960): ○ = endosulfan and □ = control.

maximum exists in the excitation phase II, with a subsequent decline to almost zero (Warburg vessels with 10 animals each) (Goesswald 1962; Fig. 33).

An observation of the O_2 consumption with different residence periods revealed also in *Lymantria dispar* L. larvae (L 5) (intoxication of filter paper with a dose of 100 mg of endosulfan therein) a distinct rise after a 14-hr intoxication period. In *Diprion pini* L. (larvae L 4) the O_2-consumption pattern, after an intoxication with an 0.01% aquious emulsion of endosulfan, tallied with the curve graphs for *Sitophilus* (Schulze 1965).

2. **Inner body temperature.**—Studies concerning the behavior of inner body temperature following intoxication with endosulfan were carried out by Goesswald (1962) and Schulze (1965 and 1967) with *Periplaneta americana* L.

The authors used a test arrangement in which a minute thermoelement was introduced into the lumen of the second thoracic stigma and the temperature difference from the ambient temperature was recorded while passing through the intoxication phases. The test animals were in the interior of a vessel with a constant temperature.

Schulze (1967) distinguished two reaction types:

Reaction type (a): During the latent period, scarcely any mobility is observed; the inner body temperature is almost always the same as in the untreated

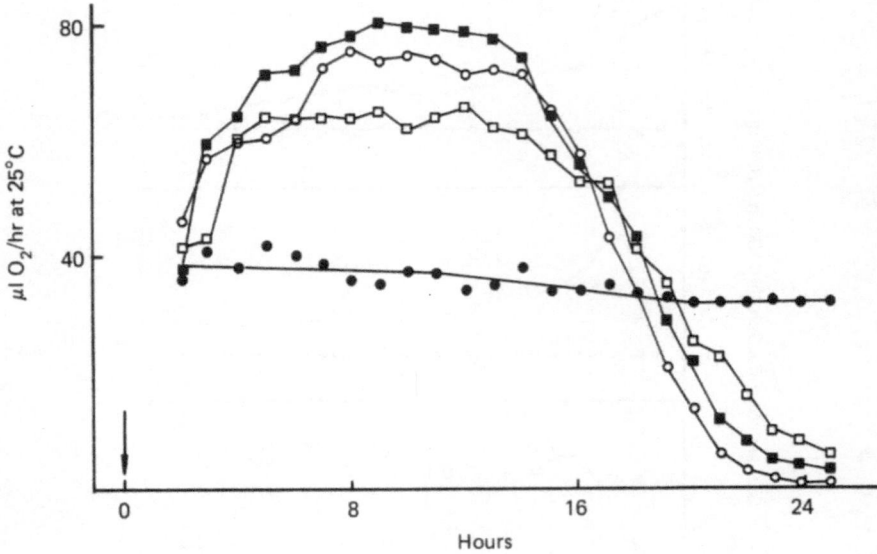

Fig. 33. Course of O_2 intake by *Sitophilus granarius* L. after intoxication with
endosulfan αI (■), endosulfan βI (□), and isomer mixtures thereof (○),
at 25°C in a long-term test (Gösswald 1962); ● = control.

controls; with the beginning of the excitation phase II there is a strong temperature rise up to 4°C over the ambient temperature; a slow decline follows.

Reaction type (b): During the latent period and the excitation phase I the
animals already show major mobile activity (temperature peaks 1.4 to 4.2°C
over the ambient temperature). Excitation phase II is reached earlier as in the
case of reaction type (a); here the body temperatures are maximally only 1.4°C
above the ambient temperature (Fig. 34).

The rise of inner body temperature is accounted for as an indirect result of
endosulfan intoxication, due to increased muscle activity. One can see that a
temperature rise in the inner body invariably results from exterior activity.

3. Effect on water turnover.—The course of transpiration in *Blaberus trapezoides* Burm. was studied by Schulze (1965 and 1967) by means of tritiated
water synchronically with the temperature course. Prior to the experiment the
test animals received an injection of 0.05 ml THO (corresponding to 0.1 mC)
directly into the haemolymph. After 48 hr the animals were transferred to the
test chamber and set on 30-cm² filter paper sheets containing 200 mg of endosulfan (applied as acetone solution). By a constant and controlled air stream the
respiration water was conducted to a gas-flow ionization chamber. The temperature in the chamber holding the animals was kept constant at 28°C, and at
constant air humidity (100%, 68%, and 20% relative atmospheric humidity) over
a period of 24 hr, the occurring radioactivity was measured. It was found that

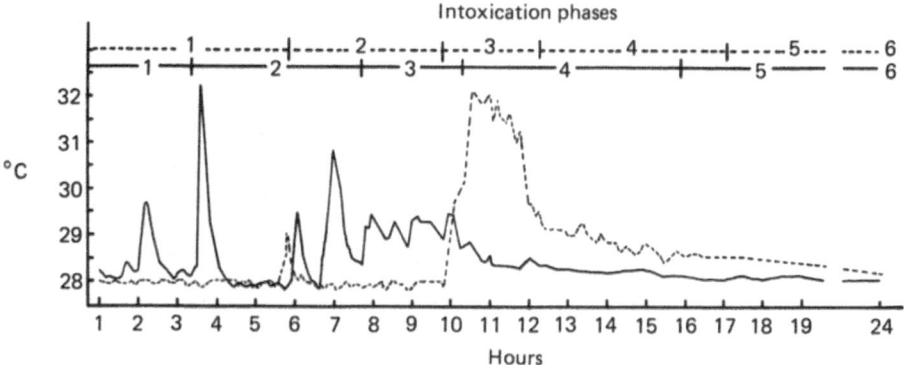

Fig. 34. Different reaction types of body temperature course demonstrated according to results at 28°C and 54% relative atmospheric humidity (Schulze 1967): - - - - a and ——— b.

the transpiration curves are broadly parallel to the body temperature curve (Fig. 35).

Observations with cockroaches (*Blaberus trapezoides* Burm.) show that they lose weight partly to such an extent that during preparation of the insects killed by endosulfan their body cavity appears to be dried out (Schulze 1965).

4. **Pulse rate.**—Schulze (1965) investigated pulse rate alterations in *Diprion pini* L. (L 4) after a treatment with endosulfan. The caterpillars were intoxicated by dipping them into a 3% ethanolic solution of endosulfan of the concentrations 0.1 and 0.01% a.i. (dipping time 5 sec). After short drying the animals were

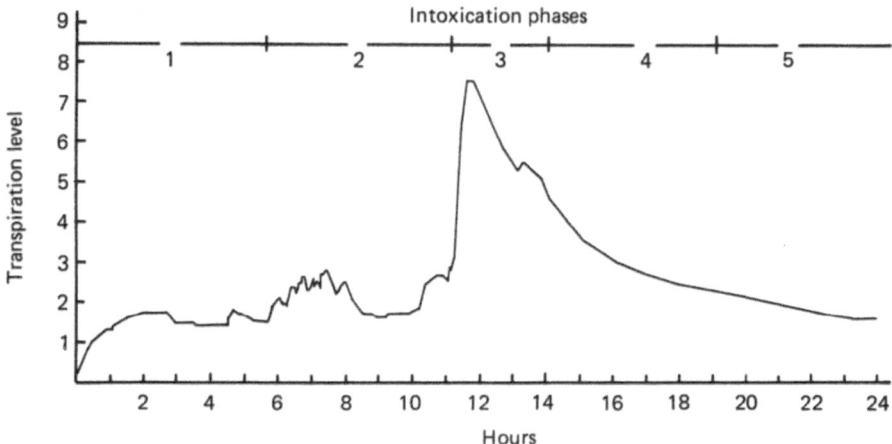

Fig. 35. Transpiration course of intoxicated male *Blaberus trapezoides* Burm. at 28°C and 68% relative atmospheric humidity (Schulze 1967).

observed under a binocular microscope and the stroke frequency of the dorsal vessel counted using a stopwatch. It was found that increasing excitation of the animals under test is concomitamt with an increase of pulse rate which reaches its culmination in the excitation phase III with 110 to 130 strokes/min. (test temperature 25°C). Subsequently there is a decline of frequency, the tremor phase showing irregular pulse rate, finally leading to complete stoppage (Fig. 36).

These results tally with investigations in the water flea (*Daphnia pulex* O. F. Müll.); here, too, a decline of pulse rate can be observed with progressive poisoning and eventually intermittent failure (*Hoechst AG* 1970).

5. pH value of hemolymph.– Schulze (1965) could observe a permanent decline of said pH value during the intoxication phases in *Diprion pini* L. by means of a pH meter and the "one-drop electrode assembly" technique (Fig. 37). Such decline is also evidenced in an intoxication with phosphoric acid esters. The causes of progressing increase of acidity in the hemolymph are not known so far.

6. Further reactions.–In the course of intoxication of insects with endosulfan, processes can be recorded that also occur with other plant pesticides. The animals (*e.g., Periplaneta americana* L. and *Blaberus trapezoides* Burm.) inflate gradually during the tremor phase. During preparation of the abdomen it becomes noticeable that the intestinal lumen is filled with air. It is assumed that this effect is due to prolonged convulsive swallowing of air (Schulze 1965).

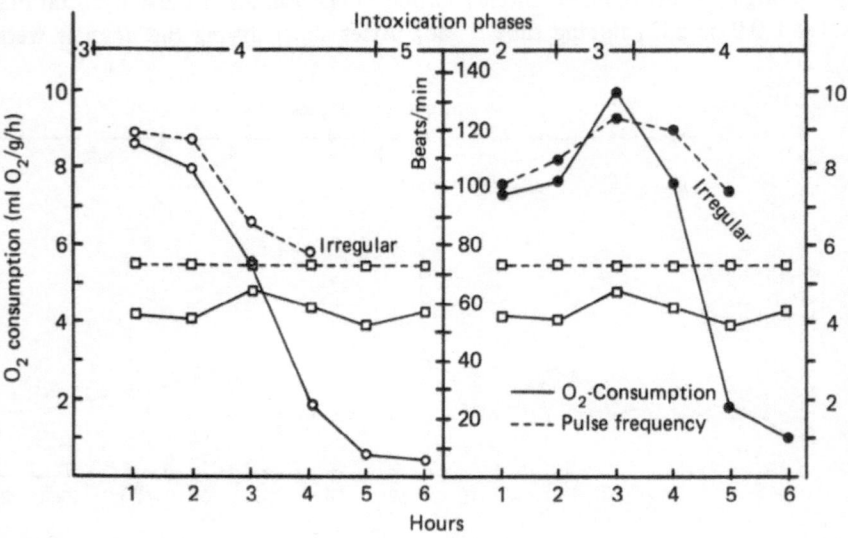

Fig. 36. O_2 consumption and pulse rate of *Diprion pini* L. (IVth instar) at 25°C (Schulze 1965): ○ = endosulfan 0.1%, ● = endosulfan 0.01%, and □ = control.

Goesswald (1958) observed histopathologic alterations especially in the region of ganglion cells. Such pyknoses of the nuclei become visible after dyeing of sections of the termite *Calotermes flavicollis* F. with gallocyanin and counter-dyeing with eosin, especially in the thoracic ganglia. For this purpose the animals were kept for 48 hr on round filter papers containing each 6 ml of a 1% solution of endosulfan emulsion and subsequently fixed in a Karl-mixture.

Klee (1960) was able to confirm this on nuclei of motoric neurons of the thoracic ganglia of *Calotermes*. He could further observe, after intoxication with endosulfan and the reaching of its "anesthetic phase", vacuole formation in the cytoplasm of the central intestinal cells. The chromatin of the cell nuclei showed intensive clotting.

Summary

Endosulfan acts on arthropods as a contact and feed (stomach) poison. An effect over the gas phase is only encountered in closed rooms; it is then considered as indirect contact effect caused by precipitation of the sublimated active substance on the integument of the insect.

The effect on insects of many orders and families is evidenced both as a contact and as a feed poison.

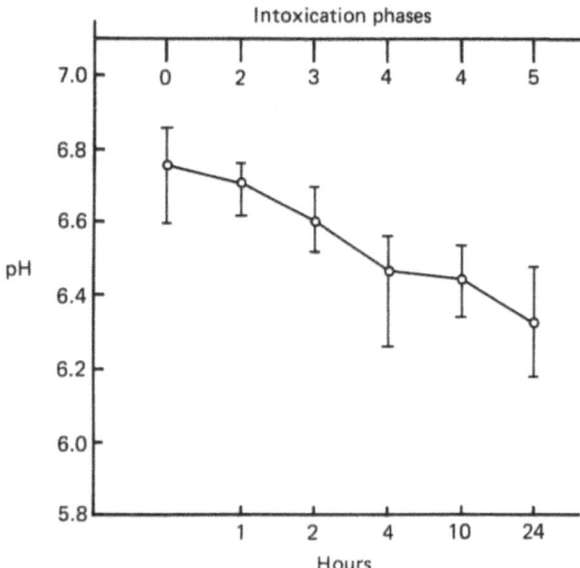

Fig. 37. Decrease of pH value during intoxication course following Thiodan poisoning: C = control at room temperature and atmospheric humidity (Schulze 1965).

Certain intoxication symptoms in insects are typical for endosulfan; in the process of intoxication the active substance is regularly taken up and dispersed over the haemolymph. The effect of endosulfan starts in the centers of motor stimulation; cell injuries can also be evidenced there. During intoxication, the activity of the insects is increased in certain phases, expressed by a spontaneous rise of body temperature as well as of oxgen consumption and release of water vapor. Of the two endosulfan isomers, the α isomer rather than the β isomer was found to be the more potent insecticide. This is evidently accounted for by a difference in the rate of penetration in to the insect; endosulfan sulfate (IV) is equally toxic to insects. As for the metabolites endosulfan diol (II), endosulfan ether (III), and hydroxyendosulfan ether (V), no toxic effect against insects has been ascertained so far.

VIII. Tolerances for endosulfan

By R. H. Rimpau

Since the publication of Maier-Bode (1968) many endosulfan tolerances have been issued for various crops in the Federal Republic of Germany, Australia, Canada, New Zealand, The Netherlands, Belgium, France, Austria, South Africa, and by FAO/WHO. Table XLVIII is a survey thereof.

In 1969, the endosulfan tolerances for meat, fat, and meat products of slaughter cattle were raised to 0.2 mg/kg, for milk fat to 0.5 mg/kg. For endosulfan in cottonseeds, a tolerance of 1.0 mg/kg was established in the U.S.A.; this tolerance was recognized and accepted by FAO/WHO in 1974. Table XLVIII does not cover all pertinent countries; on the strength of a few examples the development of endosulfan tolerances is merely illustrated. Table XLIX presents a separate survey of the tolerances granted in the U.S.A. (as of December 1, 1978).

IX. Thiodan[®]: Fields of application and survey of technical literature.

By H. Hüttenbach

a) Introduction

The insecticide Thiodan[®] (active ingredient endosulfan) has been in use for some years now in most countries of the world. Numerous trials have been carried out with this product, which is the subject of many scientific publications. In this section an attempt is made to survey the present applications of Thiodan.

Thiodan acts on arthropods as a contact and stomach poison and at higher temperatures also as a respiratory poison via its gas phase. It controls chewing and sucking insects as well as mites of the families *Eriophyidae* and *Tarsonemidae*. Thiodan protects treated plants for many days against re-infestation. Experience has shown that its behavior with regard to biocenosis is very favorable. It is very well tolerated by plants and in many cases has given rise to higher yields which cannot be exclusively attributed to its insecticidal properties.

Thiodan does not accumulate in the organs of warm-blooded animals. It is not persistent in nature in that it does not form residues which remain active for unacceptably long periods.

Owing to these properties Thiodan is now used as an insecticide for protecting a wide and varied range of crops. Details of various formulations of Thiodan and of its most important applications are given on the following pages.

b) Formulations

Thiodan dust formulations containing 1,[1] 3, or 4 % technical grade endosulfan are suitable for ground and aerial application. For aerial application a dust formulation with a higher bulk density is used in order to reduce drifting. In recent years the use of insecticidal dust formulations has markedly decreased and granular and liquid formulations are now generally preferred.

Thiodan granules containing 1,[1] 3, 4, or 5 % technical grade endosulfan can be applied by hand, by mechanized ground equipment, or from the air against stem borers in paddy rice, sorghum, maize, and sugarcane. In rice Thiodan granules are applied to the standing water in paddy fields. In maize, sorghum, and sugarcane, stem borers are controlled by foliar applications.

Thiodan wettable powder containing 35 or 50% technical grade endosulfan can be applied by spraying or atomizing from the air by plane or helicopter but is intended primarily for application by mechanized ground equipment.

Thiodan emulsifiable concentrate containing 35% technical grade endosulfan can be applied by spraying or atomizing like Thiodan wettable powder. However, it is more suitable than the latter for aerial application since there is no risk of nozzles becoming clogged even if a minimum of water is used.

Thiodan ULV containing 25% technical grade endosulfan is specifically intended for ultra-low-volume (ULV) application either by mechanized ground equipment or from aircraft. This formulation allows the active ingredient to be applied in the desired spectrum of 50 to 100 microns diameter without the risk of evaporation. No further additives are required.

c) Application rates, special notes, and tolerances

The required application rate depends on the type of insect pest, its stage of development, as well as on the height and density of the crop to be treated. It generally ranges between 300 and 1,100 g a.i./ha. Application conditions vary considerably on a world-wide basis. The concentrations or application rates given in Section IV are, therefore, intended for guidance only. They may be increased or reduced as appropriate to the circumstances. Dust formulations are generally

---

[1] Only for controlling *Busseola fusca*.

Table XLVIII. Endosulfan tolerances (Endosulfan -α, -β, and -sulfate).

Crop	German F.R.	FAO	Australia	Canada[a]	New Zealand	The Netherlands[b]	Belgium	France	Austria	South Africa[a]
Almonds	(1.0)	(2.0)	(1.0)		(2.0)	(0.5)	(0.5)	(0.5)	(0.5)	0.5
Apples	(1.0)	(2.0)	(1.0)	2.0	(2.0)	(0.5)	(0.5)	(0.5)	(0.5)	
Apricots	(1.0)	(2.0)	(1.0)	2.0	(2.0)	(0.5)	(0.5)	(0.5)	(0.5)	
Artichokes	(1.0)	(2.0)	(1.0)	1.0	(2.0)	(0.5)	(0.5)	(0.5)	(0.5)	
Asparagus	(1.0)	(2.0)	(1.0)		(2.0)	(0.5)	(0.5)	(0.5)	(0.5)	
Avocadoes	(1.0)	(2.0)	(1.0)		(2.0)	(0.5)	(0.5)	(0.5)	(0.5)	
Bananas	(1.0)	(2.0)	(1.0)			(0.5)	(0.5)	(0.5)	(0.5)	
Barley grains	(0.1)		(1.0)			(0.5)	(0.5)	(0.5)	(0.1)	
Beans (Phaseolus spp.)	(1.0)	(2.0)	(1.0)	1.0	(2.0)	(0.5)	(0.5)	(0.5)	(0.5)	2.0
Blackberries	(1.0)	(2.0)	(1.0)		(2.0)	(0.5)	(0.5)	(0.5)	(0.5)	
Black salsify	(1.0)	(2.0)	(1.0)		(2.0)	(0.5)	(0.5)	(0.5)	(0.5)	
Blueberries	(1.0)	(2.0)	(1.0)		(2.0)	(0.5)	(0.5)	(0.5)	(0.5)	
Broad beans	(1.0)	(2.0)	(1.0)		(2.0)	(0.5)	(0.5)	(0.5)	(0.5)	
Broccoli	(1.0)	(2.0)	(1.0)	2.0	(2.0)	(0.5)	(0.5)	(0.5)	(0.5)	(2.0)
Brussels sprouts	(1.0)	(2.0)	(1.0)	2.0	(2.0)	(0.5)	(0.5)	(0.5)	(0.5)	(2.0)
Cabbage	(1.0)	(2.0)	(1.0)	2.0	(2.0)	(0.5)	(0.5)	(0.5)	(0.5)	2.0
Carrots	0.2	0.2	(1.0)		(2.0)	(0.5)	(0.5)	0.2	0.2	
Cashew nuts	(1.0)	(2.0)	(1.0)		(2.0)	(0.5)	(0.5)	(0.5)	(0.5)	
Cauliflower	(1.0)	(2.0)	(1.0)	1.0	(2.0)	(0.5)	(0.5)	(0.5)	(0.5)	(2.0)
Celery (whole)	(1.0)	(2.0)	(1.0)	1.0	(2.0)	(0.5)	(0.5)	(0.5)	(0.5)	
Collards	(1.0)	(2.0)	(1.0)	(2.0)	(0.5)	(0.5)	(0.5)	(0.5)	(0.5)	(2.0)
Champignons	(1.0)					1.0	1.0			
Chard	(1.0)	(2.0)	(1.0)		(2.0)	(0.5)	(0.5)	(0.5)	(0.5)	
Cherries	(1.0)	(2.0)	(1.0)	2.0	(2.0)	(0.5)	(0.5)	(0.5)	(0.5)	
Chestnuts	(1.0)	(2.0)	(1.0)		(2.0)	(0.5)	(0.5)	(0.5)	(0.5)	

Chicory	(1.0)	(2.0)	(1.0)		(2.0)	(0.5)	(0.5)	(0.5)	
Citrus	(1.0)	(2.0)	(1.0)		(2.0)	(0.5)	(0.5)	(0.5)	2.0
Coconuts[a]	(1.0)	(2.0)	(1.0)		(2.0)	(0.5)	(0.5)	(0.5)	
Cottonseed		1.0	1.0					(0.2)	
Cottonseed oil		0.4						(0.1)	
Cucumbers	(1.0)	(2.0)	(1.0)	1.0	(2.0)	(0.5)	(0.5)	(0.5)	(2.0)
Currants	(1.0)	(2.0)	(1.0)		(2.0)	1.0	1.0	(0.5)	
Dates	(1.0)	(2.0)	(1.0)		(2.0)	(0.5)	(0.5)	(0.5)	
Eggplant	(1.0)	(2.0)	(1.0)	1.0	(2.0)	(0.5)	(0.5)	(0.5)	
Flax seeds		(2.0)	(1.0)					(0.2)	
Fruit	1.0	2.0	1.0		2.0	0.5	0.5	0.5	
Fennel greens, vegetable	(1.0)	(2.0)	(1.0)	(2.0)	(2.0)	(0.5)	(0.5)	(0.5)	
Garlic	(1.0)	0.2	(1.0)		(2.0)	(0.5)	(0.5)	(0.5)	
Gooseberries	(1.0)	(2.0)	(1.0)		(2.0)	(0.5)	(0.5)	(0.5)	
Grapes	(1.0)	(2.0)	(1.0)	1.0	(2.0)	(0.5)	(0.5)	(0.5)	
Green beans	(1.0)	(2.0)	(1.0)		(2.0)	(0.5)	(0.5)	(0.5)	2.0
Hazelnuts	(1.0)	(2.0)	(1.0)		(2.0)	(0.5)	(0.5)	(0.5)	
Horseradish	(1.0)	(2.0)	(1.0)		(2.0)	(0.5)	(0.5)	(0.5)	
Kale	(0.1)	(2.0)	(1.0)		(2.0)	(0.5)	(0.5)	(0.5)	
Leek	(1.0)	(2.0)	(1.0)		(2.0)	(0.5)	(0.5)	(0.5)	(2.0)
Lettuce	(1.0)	(2.0)	(1.0)	2.0	(2.0)	(0.5)	(0.5)	(0.5)	
Macadamia nuts	(1.0)	(2.0)	(1.0)		(2.0)	(0.5)	(0.5)	(0.5)	
Maize	0.15							0.15	
Mangoes	(1.0)	(2.0)	(1.0)		(2.0)	(0.5)	(0.5)	(0.5)	
Meat fat		0.2	0.2[a]						
Melons	(1.0)	(2.0)	(1.0)	1.0	(2.0)	(0.5)	(0.5)	(0.5)	(2.0)
Milk		0.5	(1.0)						
Milk fat			0.5[b]						

Table XLVIII (continued)

Crop	German F.R.	FAO	Australia	Canada[a]	New Zealand	The Netherlands[b]	Belgium	France	Austria	South Africa[a]
Millet	(0.1)		(1.0)						(0.1)	
Mushrooms	(1.0)					1.0	1.0			
Mustard seeds									(0.2)	
Mustard greens	(1.0)	(2.0)	(1.0)		(2.0)	(0.5)	(0.5)		(0.5)	
Nectarines	(1.0)	(2.0)	(1.0)	2.0	(2.0)	(0.5)	(0.5)	(0.5)	(0.5)	
Oats grains	(0.1)		(1.0)					(0.5)	(0.1)	
Oil-bearing seeds	(1.0)									
Olives	(1.0)		(1.0)						0.2	
Onions	(1.0)	0.2	(1.0)		(2.0)	(0.5)	(0.5)	(0.5)	(0.5)	
Paprika	(1.0)	(2.0)	(1.0)	1.0	(2.0)	(0.5)	(0.5)	(0.5)	(0.5)	
Parsley	(1.0)	(2.0)	(1.0)		(2.0)	(0.5)	(0.5)	(0.5)	(0.5)	
Peas, green	(1.0)	(2.0)	(1.0)	0.5 ("peas")	(2.0)	(0.5)	(0.5)	(0.5)	(0.5)	2.0
Peaches	(1.0)	(2.0)	(1.0)	2.0	(2.0)	(0.5)	(0.5)	(0.5)	(0.5)	0.5
Pears	(1.0)	(2.0)	(1.0)	2.0	(2.0)	(0.5)	(0.5)	(0.5)	(0.5)	0.5
Peanuts[c]	(1.0)	(2.0)	(1.0)		(2.0)	(0.5)	(0.5)	(0.5)	(0.5)	
Pecan nuts	(1.0)	(2.0)	(1.0)		(2.0)	(0.5)	(0.5)	(0.5)	(0.5)	
Pineapples, fruit	(1.0)	(2.0)	(1.0)		(2.0)	(0.5)	(0.5)	(0.5)	(0.5)	
Pistachio	(1.0)	(2.0)	(1.0)		(2.0)	(0.5)	(0.5)	(0.5)	(0.5)	
Plums	(1.0)	(2.0)	(1.0)	2.0	(2.0)	(0.5)	(0.5)	(0.5)	(0.5)	
Poppy seeds	(0.1)								(0.2)	
Potatoes	(1.0)	0.2				0.1	(0.5)			
Prunes	(1.0)	(2.0)	(1.0)	2.0	(2.0)	(0.5)	(0.5)	(0.5)	(0.5)	
Pumpkins and squash	(1.0)	(2.0)	(0.1)	1.0	(2.0)	(0.5)	(0.5)	(0.5)	(0.5)	
Radish	(1.0)	(2.0)	(1.0)		(2.0)	(0.5)	(0.5)	(0.5)	(0.5)	(2.0)

Crop									
Red beets	(1.0)	(2.0)	(1.0)		(2.0)	(0.5)	(0.5)	(0.5)	
Red radish	(1.0)	(2.0)	(1.0)		(2.0)	(0.5)	(0.5)	(0.5)	
Rape, *Brassica napus oleifera*	0.5							(0.2) (Seed)	
Rape-seed, *Brassica rapa*	0.5								
Rhubarb	(1.0)	(2.0)	(1.0)		(2.0)	(0.5)		(0.2)	
Rice	(0.1)	0.1[d]					(0.5)	(0.5)	
Rye grains	(0.1)		(1.0)					(0.1)	
Safflower								(0.1)	0.2
Spices								(0.1)	
Spinach	(1.0)	(2.0)	(1.0)	2.0	(2.0)	0.5	(0.5)	(0.5)	
Strawberries	(1.0)	(2.0)	(1.0)	1.0	(2.0)	(0.5)	(0.5)	(0.5)	
Sugarbeet root	(0.1)							(0.1)	
Sugarcane	(0.1)							(0.1)	
Sunflower seeds								(0.2)	
Sweet corn	(1.0)					(0.5)	(0.5)	(0.5)	
Sweet potatoes	(0.1)	0.2							
Tea	30.0[e]	30.0[e]						(0.1)	
Tomatoes	(1.0)	(2.0)	(1.0)	1.0	(2.0)	(0.5)	(0.5)	(0.5)	2.0
Turnip, garden	(1.0)	(2.0)	(1.0)		(2.0)	(0.5)	(0.5)	(0.5)	
Turnip greens	(1.0)	(2.0)	(1.0)		(2.0)	(0.5)	(0.5)	(0.5)	
Vegetables	1.0	2.0	1.0		2.0	0.5	0.5	0.5	
Walnuts	(1.0)	(2.0)	(1.0)		(2.0)	(0.5)	(0.5)	(0.5)	
Watercress	(1.0)	(2.0)	(1.0)	1.0	(2.0)	(0.5)	(0.5)	(0.5)	
Wheat	(0.1)		(1.0)					(0.1)	
Youngberries									2.0

[a] Without endosulfan sulfate. [b] Without specifying crops concerned. [c] Not as oil seed. [d] In husk. [e] Dried.

Table XLIX. *Tolerances for endosulfan in the U.S.A.*[a]

0.3	Alfalfa (fresh)
1	Alfalfa hay
0.2 (N)	Almonds
1	Almond hulls
2	Apples
2	Apricots
2	Artichokes
0.1 (N)	Barley grain
0.2 (N)	Barley straw
2	Beans
0.1 (N)	Blueberries
2	Broccoli
2	Brussels sprouts
2	Cabbage
0.2	Carrots
0.2	Cattle (meat, fat, meat by-products)
2	Cauliflower
2	Celery
2	Cherries
2	Collards
0.2	Corn, sweet (kernels plus cobs)
1	Cottonseed
2	Cucumbers
2	Egg plants
0.2 (N)	Filberts
0.2	Goats (meat, fat, meat by-products)
2	Grapes
0.2	Hogs (meat, fat, meat by-products)
0.2	Horses (meat, fat, meat by-products)
2	Kale
2	Lettuce
0.2 (N)	Macadamia nuts
2	Melons
0.5 (N)	Milk fat
2	Mustard greens
0.2 (N)	Mustard seed
2	Nectarines
0.1 (N)	Oat, grain
0.2 (N)	Oat, straw
2	Peaches
2	Pears
2	Peas (succulent type)
0.2 (N)	Pecans
2	Peppers
2	Pineapples
2	Plums

Table XLIX (*continued*)

0.2 (N)	Potatoes
2	Prunes
2	Pumpkins
0.2 (N)	Rape seed
0.1 (N)	Rye, grain
0.2 (N)	Rye, straw
0.2 (N)	Safflower seed
0.2	Sheep (meat, fat, meat by-products)
2	Spinach
2	Strawberries
0.1 (N)	Sugarbeets (without tops)
0.5	Sugarcane
2	Summer squash
2	Sunflower seed
0.2	Sweet potatoes
24 FA	Tea, dried
2	Tomatoes
2	Turnip greens
0.2 (N)	Walnuts
0.2 (N)	Watercress
0.1 (N)	Wheat, grain
0.2 (N)	Wheat, straw
2	Winter squash

[a] (N) = negligible residue tolerance and FA = food additive tolerance.

applied at a 10 to 20% higher rate of a.i. than are liquid formulations. The given concentrations relate to the high-volume method of application. For low-volume application, the concentration must be increased accordingly.

With certain exceptions such as the anti-termite treatment of soil or plant sets, it is generally recommended to start treatment as soon as the first signs of infestation are observed. Caterpillars should be controlled at an early stage of development, since later larval stages are considerably more resistant. It depends on the severity and duration of an infestation, whether or not the treatment has to be repeated at intervals of 10 to 14 days.

Although Thiodan is not recommended for controlling mites of the *Tetranychidae* family it is fairly effective against them, especially at higher temperature. Reynolds *et al.* (1969) confirmed this for *T. cinnabarinus*. In two other publications this acaricidal action of Thiodan is mentioned, but in relation to different crops. Pierza and Fisher (1965) made this observation with *Panonychus citri* and Young and Ditman (1959) with *Tetanychus cinnabarinus*.

Thiodan possesses limited penetration properties only. Hence, treatment of the affected side of plants is essential to achieve good control. This is particularly important when resistance to phosphoric acid esters in aphids necessitates a change to Thiodan, which is fully effective also under these conditions.

WHO/FAO have stipulated the following tolerances for endosulfan which take into account its toxicological behavior and practical plant protection requirements:

TOLERANCES

Tea (dry, manufactured)	30.0 mg/kg
Fruits	2.0 mg/kg
Vegetables (other than exceptions noted)	2.0 mg/kg
Carrots, potatoes, sweet potatoes, bulb onions	0.2 mg/kg
Cottonseed	1.0 mg/kg
Cottonseed oil (crude)	0.4 mg/kg
Rice, in husk	0.1 mg/kg
Milk and milk products (fat basis)	0.5 mg/kg
Fat of meat	0.2 mg/kg

d) Fields of application

1. Annual field crops: Grain crops.—

α) *Wheat.*—In subtropical areas devastating damage is sometimes caused by the caterpillar of the noctoid *Spodoptera exempta*. About 2.0 L/ha Thiodan 35 EC (approximately 700 g a.i./ha) is applied to control this pest (Anonymous 3, 1965; Bot *et al.* 1973).

Thiodan has proved very effective against the cereal bug *Aelia rostrata*, which frequently occurs in large numbers in the western Mediterranean. In Spain, excellent results have been obtained with Thiodan 35 EC and Thiodan ULV (800 to 900 g a.i./ha) (Anonymous 12, 1971).

In France, an insecticide-fungicide combination has been successfully applied for some years now as a seed dressing at a rate of 600 g of product/100 kg of seed. The insecticidal components of this product are 33.5 % Thiodan technical plus 13.3 % lindane. It protects grain crops from damage by the following insect pests: *Agriotes* sp., *Oscinella frit*, and *Hylemya coarctata*. The latter pest is mainly controlled by the Thiodan component.

β) *Maize (corn) and sorghum.*—Stem borers *Ostrinia nubilalis* (Engel 1971 and 1972, Kaplan-Reiterer 1971, Mustea *et al.* 1970), *Diatraea* sp. (Henderson and Davis 1970, Randolph *et al.* 1967), *Chilo zonellus* (Anonymous 9, 1970; Gangaprasada 1970), *Busseola fusca* (Bot *et al.* 1973) are the most important of the pests encountered in these crops. Thiodan EC or WP is applied at a rate of 500 to 800 g a.i./ha to control them. In many countries, Thiodan granules, applied at approximately the same rate, are preferred. The granules roll into the leaf axils and whorls where they come into contact with the young caterpillars and kill them before they can penetrate the plant. In general, the 3% granular formulation is applied, but the 1% granular formulation is also used to control *Busseola fusca*. In the treatment of individual plants, the application rate is much lower. Good results have also been obtained with Thiodan ULV, especially in the control of *Busseola fusca*. The following authors among others deal with the control of stem borers: Anonymous 9 (1970), Engel (1971 and 1972), Gangaprasada (1970), Henderson and Davis (1970), Kaplan-Reiterer (1971), Mustea *et al.* (1970), and Randolph *et al.* (1967).

Leaf-eating caterpillars such as *Spodoptera* sp. and *Cirphis unipuncta* are controlled by Thiodan EC or WP applied at the above rate.

The caterpillars of *Heliothis* which cause considerable damage by invading ears are controlled by dusting or spraying with Thiodan (Anderson *et al.* 1960, Rawat *et al.* 1970). In individual plant treatment the female inflorescence is dusted. A much lower dose is sufficient under these conditions.

To control the gall midge *Contarinia* sp., an application rate of 2.0 L of Thiodan 35 EC/ha is recommended, although good results have also been obtained with as little as 1.0 L/ha (Anonymous 9, 1970, Gangaprasada 1970).

Spraying with Thiodan has proved effective against the leafhopper *Peregrinus maidis* (Rathore *et al.* 1970) and the aphid species *Rhopalosiphum maidis* and *Schizaphis graminum* (Depew 1971).

To control cutworms (*Agrotis* sp., *Euxoa* sp.) in maize and many other crops, the soil surface and young plants should be sprayed with Thiodan WP (700 to 1,000 g a.i./ha) or treated with Thiodan dust.

In some countries the application of 30 to 40 kg of bait/ha (Bot *et al.* 1973) is preferred. The bait is prepared as follows: 100 kg of a suitable bait material (*e.g.*, wheat bran or maize meal) is mixed with 300 to 500 g a.i. in the form of Thiodan EC or WP. Immediately before distributing the bait in the evening, it is moistened to form a crumbly mass, which is broadcast over the affected area.

γ) *Rice*.—In rice crops two formulations are widely used, *i.e.*, Thiodan 35 EC and Thiodan 5% granules.

Thiodan 35 EC applied at a rate of 1.5 to 2.0 L/ha controls stem borers *Chilo* sp., *Chilotraea* sp., *Sesamia* sp., and *Tryporyza* sp. (Anonymous 5, 1967, Calora and Ferrino 1964, Chatterjee 1972, Mehesare 1970, Pathak 1968), leaf-eating caterpillars *Cirphis* sp., *Cnaphalocrocis medinalis* (Alam 1965, Mehesare 1970), *Nymphula depunctalis* (Mehesare 1970, Srivastava *et al.* 1970), *Pseudaletia separata* (Mehesare 1970, Purohit *et al.* 1971), *Spodoptera* sp. (Mehesare 1970), leaf hoppers *Nephotettix* sp. (Alam 1965, Chatterjee 1972, Mehesare 1970), *Sogatella* sp. (Mehesare 1970), bugs *Leptocorisa* sp. (Mehesare 1970), *Solubea* sp., and beetles *Hispa armigera* (Alam 1965, Mehesare 1970). The granules are applied in a concentration of 1.0 to 2.0 kg a.i./ha for controlling the above stem-borer species in paddy rice. Whereas the emulsion is applied to the plant surface and adheres to it, most of the granules fall into the water of the paddy fields, from where the active ingredient exerts its effect by capillary action between the stem and the leaf sheath and also by evaporation. The long-term effect of the granular formulation is far superior to that of the emulsifiable concentrate under these conditions.

Alam (1965) also recommends Thiodan for the control of *Dasychira securis, Leptispa pygmaea, Melanitis leda ismene, Oscinella frit, Pelopidas agna,* and *Selenocephalus virescens*.

2. Annual field crops: Leguminous crops.—

α) *Beans and peas*.—In India, good results have been obtained in the control of *Etiella zinckenella* (Anonymous 21, 1972) with as little as 1.0 L of Thiodan 35 EC/ha. The caterpillar of the polyphagous pest *Diacrisia obliqua* also occurs in these crops in India. Bakhetia and Sidhu (1970) reported on the high efficacy of Thiodan 35 EC applied at a rate of 1.2 L/ha. In earlier laboratory trials with *Diacrisia obliqua*, Pradhan *et al.* (1960) found Thiodan to be the most effective of 16 insecticides tested.

Beans of the *Vicia faba* variety are attacked by the beetle *Bruchus rufimanus*. Obarski (1969) successfully controlled this pest and at the same time increased the seed yield by the application of 1.5 L/ha of Thiodan 35 EC during flowering (main infestation period).

Thian Hua (1967) obtained very good results with Thiodan (150 g a.i./100 L of water) in the control of the bean-fly *Melanagromyza phaseoli*. The applicatons were carried out 6, 10, and 14 days after crop emergence. Lal (1971) confirmed the efficacy of Thiodan against the pupae and especially the imagos of another fly, *i.e.*, *Agromyza atricornis*, in laboratory tests.

Other insects controlled by Thiodan in these crops include beetles *Chalcodermus aeneus, Epilachna varivestis* and *Sitona lineata*; aphids *Aphis craccivora* (Bichoo 1970, Schmutterer 1969), *Aphis fabae, Acyrtosiphon pisum*; and bugs *Nezara viridula*.

Thiodan 35 EC and WP are generally applied at a rate of 300 to 500 g a.i./ha. In isolated instances crops are dusted with Thiodan dust 3%.

β) *Soybeans*.—This crop is attacked by a large number of insects against which Thiodan 35 EC applied at a rate of 175 to 525 g a.i./ha has proved effective.

In Brazil Thiodan 35 EC is officially recommended by Embrapa (1979) against the leaf-feeding caterpillars *Anticarsia gemmatalis* and *Pseudoplusia includens* at 175 g and 437 g a.i./ha, respectively, as well as against the bugs *Euschistus heros* and *Nezara viridula* at 437 g and 525 g a.i./ha, respectively. Thiodan 25 ULV is likewise recommended.

In India good results are obtained with Thiodan against the following pests: caterpillars of the butterflies *Achaea janata* (Bichoo 1970), *Anarsia ephippias* (Rawat *et al.* 1969), *Diacrisia* sp. (Bichoo 1970, Rawat *et al.* 1969, Singh 1968 b and 1969 b, Tripathi 1966), *Plusia acuta, Plusia arichalcea* (Singh 1969 b), *Spodoptera* sp. (Bichoo 1970, Rawat *et al.* 1969, Robertson 1969, Singh 1969 b and 1971); beetles *Longitarsus* sp. (Bichoo 1970), *Oberea brevis* (Bichoo 1970, Rawat *et al.* 1969), *Phyllotreta* sp. (Bichoo 1970), *Pugria* sp.; and hemipters *Acrythosiphon* sp. (Bichoo 1970), *Aphis craccivora* (Bichoo 1970, Schmutterer 1969), *Bemisia tabaci* (Singh 1968 a), *Empoasca* sp., *Myzus persicae*.

Similarly good results have been achieved in India with Thiodan in the control of the fly *Melanagromyza phaseoli*.

In Tanzania the following insects have been controlled by Thiodan: the butterflies *Heliothis armigera* and *Sylepta derogata*, as well as the bugs *Nezara viridula* and *Acanthomia* sp. (Eastern Tanzania) (Robertson 1969).

3. Annual field crops: Tuber and root crops.—

α) *Potatoes.*—Potatoes are among the first crops to have been treated with Thiodan. EC, WP, and dust formulations are applied in rates ranging from 300 to 600 g a.i./ha. Among the first to study Thiodan and its insecticidal properties outside Germany was Moore (1959) who reported on excellent results in controlling *Myzus persicae, Epitrix cucumeris, Empoasca fabae*, and *Leptinotarsa decemlineata.*

Unlike various other insecticides, Thiodan has proved to be very effective also in crops already heavily infested by aphids (Powell *et al.* 1969).

In 1972 Thiodan ULV was found to be effective in the CSSR also when applied at a rate of 600 g a.i./ha against *Leptinotarsa decemlineata* (Anonymous 22, 1972). Successful control can presumably be achieved with a much lower rate.

To control potato tuberworm *Phthorimaea operculella*, the crop must be sprayed with Thiodan at far shorter intervals than is usual. Many publications, including those of Padilla and Ortega (1964), confirm the efficacy of Thiodan against this insect.

β) *Sugarbeets.*—In this crop Thiodan is primarily used for controlling aphids (*Aphis fabae* and *Myzus persicae*). In Italy, Thiodan 35 EC is also applied at 1.5 to 1.8 L/ha against the weevil *Temnorhinus (Cleonus) mendicus* (Gelosi and Guinchi 1971) and *Lygus* sp. and at 1.2 to 1.5 L/ha against the flea beetle *Haltica* sp.

Hagen (1961) reported on the efficacy of Thiodan against *Loxostege sticticalis* and *Loxostege commixtalis.*

In Bulgaria, caterpillars of the cabbage moth *Mamestra brassicae* were successfully controlled by an aerial application of 3.0 L/ha Thiodan 35 EC (Dochkova 1969).

A combination of Thiodan and dimethoate is of interest because such a formulation simultaneously controls the beet fly *Pegomya betae*.

4. Annual field crops: Fiber and oil crops.—

α) *Cotton.*—Cotton crops are at present the most important field of application for Thiodan. Insect pests controlled by Thiodan include: leaf-eating caterpillars *Alabama argillacea, Spodoptera* sp. (Angelini and Vandamme 1965), *Syllepta derogata* and *Trichoplusia ni*; bollworms *Diparopsis* sp. (Brader 1968, Schmutterer 1972), *Earias* sp. (Ahmad and Mahsin 1969, Angelini and Vandamme 1965, Anonymous 6, 1967 and 7, 1968), *Heliothis* sp. (Ahmad and Mahsin 1969, Angelini and Vandamme 1965, Anonymous 6, 1967 and 7, 1968; Bot *et al.* 1973, Brader 1968, Davies and Ingram 1965, Fonseca 1971 and 1972, Schmutterer 1972), *Pectinophora gossypiella* (Angelini and Vandamme 1965, Sharma *et al.* 1971); cottonboll weevils *Anthonomus grandis* (Guevara

Calderon and Moreno Darme 1972), *Myllocerus maculatus* (Sharma *et al.* 1971); aphids *Aphis gossypii* (Anonymous 6, 1967 and 7, 1968, Bot *et al.* 1973, Schmutterer 1972); bugs *Helopeltis* sp. and *Lygus* sp.; leafhoppers *Empoasca* sp. (Anonymous 6, 1967 and 7, 1968, Schmutterer 1969 and 1972); thrips *Thrips tabaci*; white flies *Bemisia tabaci* (Anonymous 6, 1967 and 7, 1968, Guevara Calderon and Moreno Darma 1972, Sharma *et al.* 1971, Schmutterer 1969); and mites *Hemitarsonemus latus* (Almeida *et al.* 1966).

In the Sudan, the application of Thiodan is officially recommended at a rate of 950 g a.i./ha for controlling *Heliothis armigera, Earias insulana, Bemisia tabaci, Empoasca* sp. and *Aphis gossypii* (Anonymous 6, 1967 and 7, 1968).

After 11 years of experimental work in the Republic of Ivory Coast, Angelini and Vandamme (1965) concluded that Thiodan and endrin are the most suitable of 22 products tested for the type of cotton grown in that country. The efficacy of Thiodan against *Earias, Platyedra,* and *Spodoptera* was particularly emphasized. Schmutterer (1972) reported on the efficacy of Thiodan in controlling *Heliothis armigera, Diparopsis* sp., *Earias* sp., *Empoasca* sp., and *Aphis gossypii* in a South East African cotton plantation. In an earlier publication (1969) the same author mentioned its efficacy against *Bemisia tabaci.* In Uganda, good results were recorded in controlling *Heliothis armigera* and *Earias biplaga* (Davies *et al.* 1965). Thiodan is considered there as an alternative to DDT, which has been used for many years. Robertson (1970) arrived at similar findings in Tanzania. A subsequent report from Uganda mentioned considerable yield increases following the application of Thiodan 50 WP. The same publication also states that Thiodan proved harmless to beneficial insects. It is probably this fact which accounts for the higher yields. After alternate applications of DDT and Thiodan fewer beneficial insects were found and a lower yield was reported. The high efficiency of Thiodan ULV against *Heliothis armigera* and *Earias biplaga* is evident from the same publication (Fonseca 1971).

The application of Thiodan ULV by mechanized ground equipment or from the air has been practiced with great success for several years (Aronica 1970/71, Fonseca 1971, Watson-Cook 1973).

Ahmad and Mahsin (1969) reported from Pakistan and Aronica (1970 and 1971) from the Sudan on the successful ULV application of about 2.3 L/ha of Thiodan 35 EC against a number of pests in cotton crops. However, Thiodan 25 ULV (see above)—the formulation specially developed for this method of application—is now used with even better results.

In Chad, the sensitivity to Thiodan of the caterpillars of *Diparopsis watersi* and *Heliothis armigera*, which had become resistant to DDT and endrin, was found to be unchanged (Brader 1968).

In India, Sharma *et al.* (1971) obtained the best results with Thiodan used on its own for controlling the grey weevil *Myllocerus* sp. and the white fly *Bemisia tabaci*, whereas the combination of Thiodan and DDT was found to be most effective against the bollworms *Earias* sp. and *Pectinophora (Platyedra) gossypiella.*

The application rates for insecticides in Egypt are in general unusually high, particularly for controlling *Spodoptera littoralis*. This also transpires from the paper by Zeid *et al.* (1968) who, among others, published evidence of the remarkable ovicidal effect of Thiodan EC at 2.5 kg a.i./ha.

In Brazil, good results were obtained with 12 kg of Thiodan 3% dust/ha and 1.0 L/ha Thiodan 35 EC in controlling *Hemitarsonemus latus* (Almeida *et al.* 1964).

It is particularly important to note that Thiodan does not allow a spider mite population to develop in cotton crops. This is borne out by reports of Kock (1965), and Al-Azavi (1964) also mentioned the efficacy of Thiodan against *Earias insulana*. The control of spider mite populations (*Tetranychidae*) is due in part to the effect of Thiodan itself and partly to the fact that this insecticide is harmless to some mite predators (Reynolds *et al.* 1960, Sarospataki and Farkas 1969, Sidhu and Singh 1971). Fields treated with Thiodan throughout the growth period of cotton do normally not require the application of an additional acaricide.

In some countries Thiodan is combined with other insecticides in the following ratio of active ingredient:

> 1:0.25-0.5 triazophos
> 1:0.4-0.75 dimethoate
> 1:0.5-1.0 methylparathion
> 1:approx. 1.5 DDT:approx. 0.5 methylparathion

β) Jute.—In West Bengal, Thiodan 35 EC is applied by spraying or atomizing at a rate of 400 to 500 g a.i./ha for controlling the following pests: leaf-eating caterpillars *Anomis sabulifera, Diacrisia obliqua, Spodoptera exigua*; weevils *Apion corchori*; mites *Hemitarsonemus latus*.

γ) Rape, mustard seed, and safflower.—Thiodan EC or WP (400 to 500 g a.i./ha) is applied shortly before flowering to control the pollen beetle *Meligethes aeneus* and during flowering to control the seed weevil *Ceutorhynchus assimilis* and the pod midge *Dasyneura brassicae*. Since Thiodan is harmless to bees at the recommended dose, this treatment can also be carried out during flowering.

In the U.K., results obtained with Thiodan WP were somewhat superior to those obtained with the EC. Two applications of 500 g a.i./ha each shortly before flowering produced a slightly slower effect than comparable products but good development of seed-pods (Gould 1971).

In tests to control *Ceutorhynchus assimilis* and *Dasyneura brassicae*, Skrocki 1972) obtained the highest yield after the application of Thiodan WP.

Hornig (1962) described the full-blossom spraying of rape by helicopter with Thiodan WP at a rate of 450 g a.i. in 40 L of water/ha. Additionally three further treatments with Thiodan dust were carried out; the results were entirely satisfactory. Against *Ceutorhynchus assimilis* a long-term effect of approximately

8 days was observed, and against *Dasyneura brassicae* of at least 12 days. The experiment caused no loss of bees.

Waede (1960 and 1961) and Diehl (1961) as well as Buhl and Waede (1960) reported on the successful application of Thiodan (1,500 g a.i./ha) by the "cold fogging" technique in rape at the beginning of full flowering.

Singh (1969 a) recommended the application of Thiodan 35 EC (400 to 500 g a.i./ha) against *Achaea janata, Amsacta moorei, Antigastra catalaunalis, Diacrisia obliqua, Perigea capensis,* and the aphid *Macrosiphum sonchi.* Singh (1968 a) also achieved a high control rate of the aphid *Lipaphis erysimi* in mustard crops, using only 250 g a.i./ha as recommended.

Patel *et al.* (1971) found Thiodan 35 EC to be effective also against *Athalia lugens.* Their tests confirmed the frequently made observation that Thiodan shows higher efficacy in field application than in laboratory trials.

Prakash Sarup *et al.* (1964) and the *Extension Service of the University of California* (Anonymous 20, 1972) reported on the efficacy of Thiodan against the safflower aphid *Dactynotus carthami.*

In India the linseed midge *Dasyneura lini* belongs to the insect pests that are controlled by Thiodan EC in mustard crops (500 to 700 g a.i./ha).

δ) *Castor-oil plants.*—The dominant pests in this crop are caterpillars of the butterflies *Achaea janata* and *Euproctis lunata*; leafhoppers *Empoasca* sp., and castor leaf bugs *Eurystylus capensis.* They are controlled by spraying with Thiodan. In regions where water is not plentiful the dust formulations can be used. A rate of 500 to 900 g a.i./ha is sufficient. Weiss (1966) recommends Thiodan control of the above pests in East Africa.

ε) *Sunflower.*—The sunflower moth (*Homoeosoma electellum*) is a serious pest of this crop in North America. Carlson (1971) found that three applications of Endosulfan, 1.12 kg a.i./ha, gave effective control. Hence Thiodan became widely used in the U.S.A. for its control.

In Australia Thiodan proved to be most effective against the rutherglen bug (*Nysius vinitor*), the green vegetable bug (*Nezara viridula*), as well as against *Heliothis* sp. against which it is officially recommended.

ζ) *Peanuts.*—Aphids, apart from *Thrips* sp., *Nezara viridula, Amsacta* sp., and *Heliothis* sp. belong to the insects against which Thiodan is applied in this crop. Davies (1975) proved that Thiodan 35 EC at 490 g a.i./ha effectively controls *Aphis craccivora,* a vector of the groundnut rosette disease virus.

5. Annual field crops: Vegetable crops.—The insect pests that can be controlled with Thiodan in vegetable crops are so numerous that the examples given here are necessarily even more restricted than for other crops.

α) *Tomatoes and Spanish peppers.*—Thiodan is of special importance in these crops because it not only controls the most important pests but, owing to its residue behavior, can also be applied until one day before harvesting. Pests attacking these crops include: caterpillars of the butterflies *Heliothis* sp. (Dizon and Atienza 1971, Rolston *et al.* 1970), *Manduca sexta, Plusia acuta, Spodop-*

tera sp.; beetles *Anthonomus eugenii, Epicauta* sp., *Epitrix cucumeris, Leptinotarsa decemlineata*; aphids *Myzus persicae* (Elmore and Magor 1962, Shorey 1961); bugs *Acrosternum hilare, Nezara viridula*; white flies *Bemisia tabaci* and *Trialeurodes vaporariorum*; and mites *Vasates lycopersici*. Thiodan EC and WP at 300 to 600 g a.i./ha are applied most widely; in some countries a 3 or 4% dust formulation at 400 to 800 g a.i./ha is also used.

The greatest damage is generally caused by *Heliothis* sp. and *Myzus persicae*. Rolston *et al.* (1970) and Dizon and Atienza (1971) reported on the high efficacy of Thiodan against *Heliothis*, whereas Shorey (1961), Elmore and Magor (1962) emphasized the suitability of this insecticide for aphid control.

β) *Cabbages and onions.*—Because of the waxy plant surface of these crops the ULV application of Thiodan has proved more effective than spraying. Frequently a 3 to 4% dust formulation is also used. An application rate of 300 to 600 g a.i./ha is generally sufficient to control most of the important pests in these crops: caterpillars of the butterflies *Mamestra brassicae* (Bonnemaison 1972), *Pieris* sp. (Anonymous 19, 1972, Deshmuk and Adarsh 1971, Judge and McEwen 1970), *Plusia* sp., *Plutella maculipennis* and *Trichoplusia ni*; aphids *Brevicoryne brassicae* (McEwen *et al.* 1970), *Macrosiphum euphorbiae, Myzus persicae, Rhopalosiphum pseudobrassicae* (Anonymous 19, 1972); and thrips *Thrips tabaci*.

Young and Ditman (1959) were among the first to compare Thiodan with several other insecticides for controlling a number of vegetable pests (14 species). Thiodan and carbophenothion had the best overall effect. Judge and McEwen (1970) reported on the efficacy of Thiodan against *Pieris rapae* and *Trichoplusia ni*. The efficacy of Thiodan against these two pests and against the aforementioned aphids is further supported by the recommendations of the *Extension Service of the University of California* (Anonymous 19, 1972). However, it should be mentioned that generally higher concentrations of insecticides are applied in the U.S.A. than in most other countries.

Deshmukh and Adarsh (1971) tested the efficacy against *Pieris brassicae* under laboratory conditions. Bonnemaison (1972) described the good ovicidal and larvicidal efficiency against *Mamestra brassicae* observed in laboratory trials.

McEwen *et al.* (1970) demonstrated that a full spraying programme with Thiodan protects cabbage and cauliflower against *Brevicoryne brassicae*.

γ) *Okra (Abelmoschus).*—Insects which can be controlled with Thiodan (400 to 600 g a.i./ha) in this crop include: caterpillars of the butterflies *Earias* sp., *Heliothis armigera* (Satpathy and Mishra 1969); aphids *Aphis gossypii*; leafhoppers *Empoasca* sp. (Satpathy and Mishra 1969); and white flies *Bemisia tabaci*. Satpathy and Mishra (1969) prevented an *Empoasca* infestation for a period of 4 wk by the application of Thiodan; two spray treatments with Thiodan reduced the number of fruits damaged by *Earias* from 20.8% (untreated controls) to 8.6%.

δ) *Mushrooms.*—The mushroom flies *Phoridae* and the mushroom midges *Sciara* sp. are controlled by spraying or atomizing Thiodan WP. The first treatment is carried out immediately before spawning, the second one about 2 wk later, and the third one immediately after casing. The application rate is 100 g a.i./100 m² of bed. Woodwork, structures, etc., must also be treated, for which 50 g a.i., calculated/100 m² of bed, is additionally required (Hoechst Holland 1967).

ε) *Carrots and parsley*.—Obarski (1972) applied Thiodan 35 EC at 140 g a.i./ 100 L of water to seed stocks to control the bugs *Orthops* sp. and *Lygus* sp. Compared with the untreated controls the seed yield of carrots rose by 35% and the germination rate by 24%; in parsley the seed yield rose by 16% and the germination rate by 28%.

6. Annual field crops: Other crops.—

α) *Tobacco*.—Restrictions on the use of persistent chlorinated hydrocarbons, together with the bill enforced in the Federal Republic of Germany concerning pesticide residues in tobaccos, have increased interest in Thiodan. Moreover, the quality of tobacco is not impaired by Thiodan. The recommended dose is 275 to 550 g a.i./ha for spraying and atomizing and 300 to 600 g a.i./ha for dusting. The pests controlled are: caterpillars of the butterflies *Agrotis* sp.; *Heliothis* sp. (Anonymous 2, 1965, Bot *et al.* 1973, Mullett 1970, Patel and Patel 1969 b, Roberts and Dominick 1972, Stoeva 1968); *Manduca sexta* (Mullet 1970, Roberts and Dominick 1972); *Plusia* sp.; *Spodoptera* sp. (Jotwani *et al.* 1961, Patel and Patel 1969 a); *Trichoplusia ni*; thrips *Frankliniella fusca* (Bot *et al.* 1973), *Thrips tabaci*; aphids *Myzus persicae* (Bot *et al.* 1973, Mullet 1970, Roberts and Dominick 1972); white flies *Bemisia tabaci*; and bugs *Cyrtopeltis tenuis*.

Jotwani *et al.* (1961) as well as Patel and Patel (1969 a) reported on the efficacy of Thiodan against *Spodoptera litura*. Patel and Patel (1969 a) and Stoeva (1968) also found Thiodan to be very effective against *Heliothis armigera*. Mullett (1970) and also Roberts and Dominick (1972) recommend Thiodan for controlling *Heliothis virescens*, *Myzus persicae*, *Nezara viridula*, and *Protoparce sexta*.

In Thailand (Anonymous 2, 1965) the best results in controlling *Heliothis* were obtained with Thiodan.

Thiodan practically does not penetrate into the undamaged plant tissue. It is, therefore, necessary to ensure by selection of a suitable application method that the undersides of leaves are also treated to obtain satisfactory control of insects located there, *e.g.*, aphids.

7. Fruit crops: Soft fruit.—

α) *Currants and blackberries*.—The control of the currant gall mite *Cecidophyopsis (Phytoptus) ribis* with Thiodan has now become standard practice. Various authors deal with this subject: Clinch and Higgons (1961), van de Vrie (1967), Barth (1970), Zaets (1968). Thiodan EC or WP is applied by spraying at about 50 g a.i./100 L of water; the first application should be given at the first open flower stage, the second during full blossom, and the third at the beginning of fruit setting. This treatment also controls the black currant leaf midge *Dasyneura tetensi*, as confirmed by Michel (1968), Paetzoldt and Burmeister (1969) and van de Vrie (1967). In currants Thiodan is also effective against the aphids *Aphidula schneideri* and *Cryptomyzus ribis*.

Krczal (1969) described the damage caused by the gall mite *Eriophyes essigi* in blackberries and its control with Thiodan. The first treatment should be carried out when the new shoots are 40 cm long, the second application at a shoot length of 50 to 60 cm, and the third application during full flowering.

β) *Strawberries*.—Allen *et al.* (1957) discovered in their experiments that of 50 insecticidal substances tested, endrin and Thiodan gave the best control of the mite *Tarsonemus pallidus*. Other authors (van den Bruel 1960, Karl 1964, Mueller 1968 and 1971) confirm this advantageous property of Thiodan, whose application has now become a standard method of controlling this pest.

Prior to transplanting runners, above-ground parts including their crowns should be dipped into a suspension or emulsion of Thiodan containing 50 g a.i./100 L of water. In established crops the plants must be thoroughly sprayed with the same concentration before flowering and after harvest. 1,500-2,000 L of spray liquid/ha are recommended.

γ) *Pineapples*.—In the Philippines it was first shown that the mite *Steneotarsonemus ananas* can be controlled with Thiodan. This led to the registration of Thiodan in the U.S.A. for this indication. Le Grice and Marr (1970) (*cf.*, also Zyl 1969) found that this mite was responsible for black spot and leather bag disease of pineapple. It could be controlled by weekly applications of Thiodan WP (approximately 1.6 kg a.i./2,200 L of water/ha/spraying) during flowering and fruit setting.

δ) *Vines*.—Dieter (1962-1965) demonstrated the possibility of controlling the grape bud mite *Eriophyes* (*Phytoptus*) *vitis* by applying Thiodan in spring from the wool-stage until the vines sprout, thus preventing early infestation of the leaves.

The grape rust mites *Epitrimerus vitis* and *Phyllocoptes vitis* are controlled by spraying with Thiodan before the buds open and after the leaves have unfolded. This is mentioned by Sarospataki and Farkas (1969) and Baggiolini *et al.* (1971). To control these mites, thorough spraying with 50 to 70 g a.i./100 L of water is suggested. In this concentration Thiodan is also effective against *Sparganothis pilleriana*.

Clysia ambiguella and *Polychrosis botrana*—especially in the first generation—can be controlled by Thiodan EC or WP (70 g a.i./100 L of water), although Thiodan in this indication does not belong to the most effective products. The harmlessness of Thiodan to honeybees is a particular advantage here since it is usually applied during flowering. To control *Haltica ampelophaga*, 50 g a.i./100 L of water is generally sufficient, although a 3 to 4% dust formulation is more suitable.

8. Fruit crops: Top fruit.—

α) *Apples and pears*.—The high efficacy of Thiodan against pests of these crops was recognized very early (Finkenbrink 1956). Insects controlled by Thiodan are: mealy apple aphids *Sappaphis mali*, green apple aphids *Aphis pomi*, wooly apple aphids *Eriosoma lanigerum*, apple psyllids *Psylla mali*, pear psyllids *Psylla pyricola*, fruit tent caterpillars *Malacosoma neustria*, winter moths *Cheimatobia brumata*, apple ermine moths *Hyponomeuta malinella*, and browntail moths *Euproctis chrysorrhoea*. Rammer *et al.* (1969) have confirmed the high efficacy of Thiodan against *Aphis pomi* and also the mealy plum aphid *Hyalopterus pruni* in trials. Various authors (Emmel 1958, Waffelaert 1962, Gupta *et al.* 1969) have reported on the use of this insecticide for controlling *Eriosoma lanigerum*.

Niemoeller (1962) described the use of Thiodan dust for controlling *Cheimatobia brumata*. Zioni (1972) characterized the four species of *Psylla* which attack pears and apples and presents Thiodan as the most suitable insecticide for their control.

Hoyt (1962) carried out tests to control the mite *Vasates schlechtendali*; he mentioned the very good effect of Thiodan in pre-blossom spraying.

Thiodan EC or WP is generally sprayed at 50 g a.i./100 L of water or atomized in a corresponding higher concentration. The following pests are also controlled by this concentration: *Erannis defoliaria, Lymantria dispar, Anthonomus pomorum, Phyllobius oblongus*, and *Anuraphis roseus*.

Pezzi (1967-1972) and Pezzani and Ruffini (1971) reported on the control of the following leaf miners: *Lithocolletis blancardella* and *Cemiostoma (Leucoptera) scitella* with Thiodan WP. The treatment is primarily directed against the imagos of the first generation. This method is now standard practice in Italy. The usual application rate is 70 g a.i./100 L of water.

β) *Peaches.*—The green peach aphid *Myzus persicae* can be effectively controlled by Thiodan at the normal concentration (50 g a.i./100 L of water) for spraying and a correspondingly higher concentration for atomizing when the application is carried out before the leaves curl. Liberal spraying with 70 to 80 g a.i./100 L water of the crown of the tree is recommended to control the twig borer *Anarsia lineatella* and of the trunk and branches to control the borers *Sanninoidea exitiosa* and *Synanthedon* sp.

Summers *et al.* (1959) reported on the action against *Anarsia lineatella*. Moore (1960) was able to reduce an infestation of *Synanthedon pictipes* by 93% by 3 applications of Thiodan. Ando (1968) achieved very good control of *Synanthedon hector* with only one application but using 100 g a.i./100 L of water.

γ) *Hazelnuts and walnuts.*—The filbert bud mite *Phytoptus avellanae*, which has become a problem in almost all hazelnut growing countries, can be successfully controlled with Thiodan EC (60 to 80 g a.i./100 L of water). The first treatment is carried out in spring after sprouting but at the latest on appearance of the third leaves of the young sprouts. A total of 3 treatments at intervals of 14 to 21 days is recommended.

In Spain Thiodan 3% dust is officially accepted also for controlling *P. avellanae* (Anonymous 16, 1972). However, better results are generally obtained with the EC.

The weevil *Balaninus nucum* is controlled with Thiodan EC or WP (80 g a.i./ 100 L of water). The first treatment is carried out on appearance of the imagos and the second 10 to 14 days later. Thiodan 4% dust has also proved to be effective at 15 to 20 kg/ha.

Pesante (1961) described the biology of *Phytoptus avellanae* in Italy and outlined the possibility of its control with Thiodan EC. As a result of his trials carried out at about the same time Krczal (1963) recommended 3 to 4 applications at 45 g a.i./100 L spray liquid. In Spain, the investigation by Marti *et al.* (1965) and Marti Fabregat and Del Rivero (1966) resulted in a large-scale campaign to control *Phytoptus* with Thiodan. Barbieri (1966) compiled the most important criteria of the use of Thiodan to control *Phytoptus* in Italy.

In California Thiodan is recommended (Anonymous 23, 1973) for controlling the aphid *Chromaphis juglandicola* and the caterpillar *Schizura concinna* in walnuts.

δ) *Mangoes.*—Leafhoppers cause much damage to this crop. In India Chari *et al.* (1969) achieved good results in controlling *Idiocerus atkinsoni* with Thiodan EC which is now widely applied at a rate of 50 to 60 g a.i./100 L of water. By thorough spraying with 30 g a.i./100 L of water Yadav and Varma (1969) destroyed 97% of the mango bud mite *Aceria mangiferae*; with 100 g a.i./100 L of water, the control rate was 99%. Valmayor (1968) recommends Thiodan in the Philippines for controlling the hoppers *Idiocerus clypealis*, *Idiocerus niveosparsus*, and *Typhlocyba nigrobilineata*. Srivastava *et al.* (1973) reduced the population of the mango mealy bug *Drosicha mangiferae* by 84.10% by a spray application of Thiodan 35 EC (50 g a.i./100 L of water, 10 L/tree). In India Thiodan is also used (50 to 60 g a.i./100 L water) to control mango psyllids *Apsylla cistellata* and weevils *Eugnamptus marginatus*.

The ant *Oecophylla smaragdina* is controlled in that country by spraying with Thiodan (350 g a.i./100 L of water). The same concentration controls the termites *Microtermes* sp. and *Odontotermes* sp. To control termites, the trunks—after destruction of the earth galleries—and the soil around the trunks are thoroughly sprayed. The soil can also be treated with Thiodan dust which should be worked in 10 cm deep.

Sidhu and Singh (1971) accomplished a 70% reduction of the population of *Oligonychus mangiferus* with Thiodan applied at a rate of only 25 g a.i./100 L of water. This shows again that even *Tetranychidae* mites can be controlled to a certain degree by regular spraying with Thiodan.

ε) *Date palms.*—In Iraq, Thiodan EC at 500 to 700 g a.i./ha has proved successful in controlling the date palm froghopper *Omnatissus binotatus*. Gharib and Djazayeri (1969) achieved a 75.6% kill of *Batrachedra amydraula* with Thiodan EC.

ζ) *Citrus.*—The following pests of this crop are controlled with Thiodan EC or WP at normal application rates: caterpillars of the butterflies *Ascotis* sp., *Papilio* sp. and *Prays citri*; thrips *Scirtothrips* sp.; citrus psyllids *Diaphorina citri*;

citrus white flies and black flies *Aleurocanthus woglumi, Bemisia tabaci, Dialeurodes citri*; and aphids *Toxoptera* sp. and *Trioza erythreae*. For white and black flies control should be directed against the imagos and first larval stages.

To give long-term protection from ant infestation, the trunks of citrus trees are sprayed liberally with Thiodan WP (200 to 300 g a.i./100 L of water).

Pradhan *et al.* (1959) tested 16 insecticides for their action against *Diaphorina citri* under laboratory conditions and found endrin, parathion, and Thiodan to be the most effective. Celino and Molino (1971) have evidence of the efficacy of Thiodan on this crop in the Philippines. In a series of tests Thorbecke and Fisher (1963) observed the efficacy of Thiodan against *Heliothis armigera, Ascotis selenaria, Papilio* sp., *Toxoptera citricida, Trioza erytreae*, and *Scirtothrips aurantii*. The tests were carried out with Thiodan WP 50% and Thiodan dust 3 to 5%.

Pierza and Fisher (1965) discovered that, unlike DDT, Thiodan WP or dust did not interfere with the parasitization of the scale insect *Aonidiella aurantii* by *Aphytis africanus*. Hodgson (1970) in his integrated pest control program for citrus crops, also applied Thiodan against *Heliothis armigera, Papilio* sp., and *Toxoptera citricidus*.

In a series of trials, Sternlicht (1969) demonstrated the high efficacy of Thiodan EC in high- and low-volume applications against *Aceria sheldoni*.

9. Plantation crops: Coffee.—Control of the coffee-berry borer *Hypothenemus (Stephanoderes) hampei* with Thiodan has now become standard practice in many countries. From Angola Fonseca Ferrao (1960) reports on damage of 10 to 20% in the crop grown at an altitude of 600 m and on the efficacy of Thiodan 35 EC at only 1.7 L/ha (approximately 600 g a.i./ha); he mentioned the possibility of aerial application of the product.

Pierrard (1962) in Rwanda mentioned the efficacy of Thiodan not only against the coffee-berry borer, but also against the bugs *Antestiopsis lineaticollis* and *Habrochila ghesquierei*.

Almeida and Cavalcante (1966) described a trial in a 6-yr-old coffee plantation in Brazil in which 8 insecticides were applied twice at an interval of 15 days. The best control of *Hypothenemus hampei* was achieved with Dieldrex and Thiodan 35 EC at 2% concentration and 150 ml/plant. Boncato and Gandia (1968) obtained good results in controlling the same pest in the Philippines with 2 L/ha of Thiodan 35 EC (approximately 700 g a.i./ha). In Uganda, Thiodan proved to be the most effective of six products under test against *H. hampei* (Ingram 1965). Schmutterer (1969) also emphasized the suitability of Thiodan for controlling this pest. Reports from Guatemala on comparative trials with 16 insecticides also confirmed the superiority of Thiodan in achieving almost complete control of *H. hampei* (Hernandez Paz 1972). A later report from Guatemala deals with comparative trials with 21 insecticides of which Thiodan 35 EC and Thiodan 3% dust produced the best results (Anonymous 18, 1972).

Further very encouraging evidence of the control of *Hypothenemus hampei* comes from Brazil. Netto *et al.* (1967) applied 2.6 L/ha of Thiodan 35 EC (approximately 900 g a.i./ha) twice at an interval of 28 days. In Minas Gerais, 2.0 L Thiodan 35 EC/1,000 plants were applied 3 times at an interval of 20 days (Anonymous 24, 1973).

Wheatley (1963) mentioned the efficacy of Thiodan against *Ascotis selenaria*. Paulino (1973) *et al.* described trials involving applications of various insecticides for controlling *Perileucoptera coffeella*, in which Thiodan EC at 490 g a.i./ha produced good results.

Thiodan is now recommended at 700 to 1,000 g a.i./ha against *Hypothenemus hampei*. The first application is generally carried out at the climax of the flight of the female beetles or when berry infestation has reached 5%. An application rate of 700 g a.i./ha is recommended to control the caterpillars of *Cephonodes hylas, Dichocrocis* sp., *Epicampoptera* sp., and the bugs *Antestiopsis* sp., *Habrochila ghesquierei, Helopeltis* sp., and *Lygus* sp.

As a preventive measure against the borer *Bixadus sierricola* the trunks are sprayed liberally with Thiodan at 600 g a.i./100 L of water.

10. Plantation crops: Cocoa.—Houillier (1961 and 1962) carried out the first trials with Thiodan against the bugs *Sahlbergella singularis* and *Distantiella theobroma* and found its effect to be equal to that of endrin and lindane.

Lavabre (1971) applied a Thiodan ULV formulation containing 17.5% a.i. at 2.0 L/ha without further dilution, *viz.*, by the ULV method. Fontan and Solo knapsack sprayers fitted with ULV adaptors were used. Infestation was reduced by 93.14 to 97.86% and a good long-term action was observed. The final report of the *Institut Français du Café et du Cacao* for 1971 (Anonymous 11, 1971) gave details of experiments with the same formulation, *viz.*, Thiodan ULV, which headed the list of tested products with a 98.6% control rate.

In tests to control the bug *Bathycoelia* in Cameroon (Anonymous 10, 1968) Thiodan 35 EC was used at 360 L/ha of spray volume. A 100% kill was achieved with 380 g a.i./ha and 92% kill with 255 g a.i./ha.

In Central and South America, infection with the fungus *Ceratocystis fimbriata* is associated with infestation by *Xyleborus*. Saunders *et al.* (1967) achieved protection for 20 weeks against infestation by *X. ferrugineus* by spraying the trunks with Thiodan EC (1,000 g a.i./100 L of water).

11. Plantation crops: Tea.—Mukerjea (1969) reported on the efficacy of Thiodan against loopers *Biston suppressarius*, bugs *Helopeltis theivora*, aphids *Toxoptera aurantii*, thrips *Scirtothrips* sp., leafhoppers *Empoasca* sp., and white grubs *Holotrichia impressa*. The author pointed out that Thiodan is well tolerated by the tea plants and does not affect the taste of the tea. Under field conditions application of Thiodan is primarily directed against *Biston suppressarius* and *Helopeltis theivora*. The application rate is in the region of 450 g a.i./ha.

Caterpillars of *Euproctis* sp., *Parasa* sp., *Thosea* sp., and the mites *Calacarus carinatus, Acaphylla theae*, and *Hemitarsonemus latus* can also be controlled by this rate of Thiodan. With caterpillars it is essential that the treatment be carried out while they are in the first stages of their development (first instars).

In India termites (*Microtermes* sp., *Odontotermes* sp.) are controlled by soil treatment with Thiodan at 2.5 kg a.i./ha. To control *Holotrichia impressa,* the soil is treated with 3.5 kg a.i./ha.

12. Plantation crops: Oil palms.—The caterpillars of *Tirathaba mundella* cause damage by feeding on the blossoms and fruit clusters. Control with Thiodan should be carried out preferably during blossom-time. Chan Check Onn (1972) reported on tests with Thiodan 35 EC at 0.04% a.i. concentration and a spray volume of about 200 L/ha. Infestation could be completely stopped by 3 applications at intervals of 7 days. Normally, 2 applications are sufficient. Wood (1972), too, emphasized the suitability of Thiodan for controlling *Tirathaba mundella.* The same author also reported on the first encouraging results in controlling the caterpillars of *Thosea bisura* with Thiodan.

Thiodan 35 EC at 700 to 1,000 g a.i./ha has proved effective in controlling *Setora nitens* and *Darna trima.*

13. Plantation crops: Rubber trees.—*Hemitarsonemus latus* and *Scirtothrips dorsalis* on young leaves and especially on top shoots are controlled by two closely spaced spray applications of Thiodan 35 EC at 0.05% a.i. concentration.

14. Plantation crops: Olives.—Official tests in Spain have shown that the olive moth *(Prays oleellus)* can be effectively controlled by Thiodan dust 3% applied at a rate of 200 to 300 g/tree during blossom-time (Caballero 1971). Further tests led to the official registration of this recommendation for the 3% dust formulation (Anonymous 17, 1972). In Spain, Thiodan dust 4% (Anonymous 13, 1972) and Thiodan 35 EC (Anonymous 15, 1971) have also been registered for the control of *Liothrips oleae.* The treatment is carried out in early spring.

15. Plantation crops: Mulberry trees.—On the strength of its own trial results (Anonymous 14, 1971), The *Sericultural Experimental Station SUWON* recommends Thiodan dust 3% at a rate of 40 kg/ha against weevils *Baris deplanata* and gall midges *Diplosis (Contarinia) mori.* The treatment should be given after pruning, but before new growth appears.

16. Plantation crops: Sugarcane.—The first encouraging results in the control of the borer *Diatraea saccharalis* were obtained with Thiodan granules 5% (Long *et al.* 1959). However, the 3 to 4% granular formulation at 15 to 20 kg/ha is now used under commercial conditions besides Thiodan EC. Kalra (1970) obtained good results against the borer *Scirpophaga nivella* with Thiodan 5% granules. In India, Thiodan 35 EC (approximately 800 g a.i./ha) and Thiodan granules 4% (20 kg/ha) are also recommended against the borers *Chilotraea infuscatella, Chilo auricilia, Chilo tumidicostalis, Proceras indicus,* and *Sesamia inferens.* It is essential that the treatment against borers be carried out before the caterpillars penetrate the plant. Two to 4 repeated applications at intervals of 10 to 14 days are necessary, allowing for variations in flight and hatching times.

Pyrilla perpusilla can be controlled with Thiodan 35 EC at only 400 g a.i./ha. In the Philippines, 100% kill of the wooly aphid *Oregma lanigera* was obtained with Thiodan 35 EC (at about 35 g a.i./100 L of water) (Anonymous 4, 1965).

In Swaziland the leafhopper *Numicia viridis* is a serious danger to sugarcane. Carnegie (1971) achieved very good control of this pest with Thiodan dust 5% at about 28 kg/ha a fact that is also expressed in the official recommendation in South Africa (Bot *et al.* 1973). To prevent damage by termites (*Odontotermes* sp., *Microtermes heimi,* etc.) 50 kg of Thiodan dust 4%/ha is applied into the furrow during planting.

17. Plantation crops: Pepper.–In East Malaysia, Thiodan 35 EC is recommended at about 80 g a.i./100 L of water against bugs *Dasynus piperis* and *Diconocoris (Diplogomphus) hewitti* and weevils *Lophobaris piperis.* An application rate of 70 g a.i./100 L of water controls aphids *Toxoptera aurantii.*

18. Forage crops: Lucerne and clover.–In their trials, Buhl and Schuette (1964) observed that the imagos of the weevil *Apion flavipes* in white clover seed stocks could be effectively controlled by a single spraying with Thiodan (at about 320 g a.i./ha). The treatment was carried out after the beginning of flowering.

In Spain, Thiodan 35 EC (Anonymous 15, 1971) and Thiodan dust 4% (Anonymous 13, 1972) are officially recognized for controlling *Apion* sp. and *Phytonomus variabilis.*

Ionescu *et al.* (1966) reported from Romania on the good results obtained by spraying with Thiodan (390 g a.i./ha) against the weevil *Tychius flavus* in lucerne. The first application is carried out as soon as the imagos finish hibernation and a second one 15 to 17 days later when new growth has started.

Quintana (1968) recorded good control of *Colias lesbia* with Thiodan 35 EC (250 g a.i./ha). The aphid *Therioaphis maculata* and the bug *Nezara viridula* are also effectively controlled.

Manglitz *et al.* (1973) achieved good results with Thiodan EC (at about 1.1 kg a.i./ha) against *Euxoa auxiliaris* which causes damage in early spring.

When tested in Saudi Arabia (Abu Yaman 1970) Thiodan 35 EC proved highly effective against the larvae of *Hypera postica (Phytonomus variabilis)* at 70 g a.i./100 L of water and is recommended for controlling this pest.

e) Forestry

The excellent results achieved with Thiodan in the control of cockchafers (*Melolontha* sp.) are well-known (Engel 1965, Homonnay 1963, Varlet 1973). It is, therefore, unnecessary to deal with this recommendation for Thiodan in detail.

The efficacy of Thiodan against bark beetles is of particular interest. Allen and Rudinsky (1959) sprayed the bark of recently felled Douglas firs with Thiodan 25 WP (120 and 360 g a.i./100 L of water). This was equivalent to an a.i. concentration of 1 and 3 g/m^2 of bark, respectively. In both concentrations Thiodan prevented infestation by *Dendroctonus pseudotsugae* until the end of the test period of 19 wk. At the higher application rate 100% protection from the beetles *Gnathotrichus sulcatus, Buprestis* sp., and *Cerambycidae*, and termites was also achieved.

Rudinsky and Terriere (1959) confirmed the efficacy of Thiodan against *Dendroctonus pseudotsugae* and stated that its long-term effect on the dead material was far better than that of lindane and a number of other insecticides. A year later, Rudinski *et al.* (1960) arrived at similar conclusions and found that the insect control measures resulted in almost complete protection from fungal infection. Good results with Thiodan in controlling *Ips grandicollis* have been reported for the first time from Australia (Rimes 1959).

Becker (1964) protected unseasoned pine logs for a period of 5 mon from infestation by *Cerambycidae, Monochamus* sp., *Gnathotrichus materarius*, and bark beetles *Hylurgops pinifex, Orthotomicus caelatus, Ips calligraphus, Ips pini,* and *Pityogenes hopkinsi* by liberal spraying with Thiodan emulsion (200 g a.i./ 100 L of water). By treating the trunks with Thiodan 50 WP (250 g a.i./100 L water) Doom and Luitjes (1970) succeeded in protecting them from *Tomicus piniperda.*

The weevils *Hylobius pales* and *Pissodes approximatus* find a favorable habitat on the stumps of trees. Bliss and Kearby (1970) liberally applied Thiodan EC (390 to 780 g a.i./100 L of oil or water) to the cut surfaces of the stumps until the preparation ran off. When oil was used as the diluent the protective action against *P. approximatus* was excellent even at the lower concentration. Similarly, *H. pales* was controlled very effectively by applying the aqueous emulsion in either concentration.

The weevil *Pissodes sitchensis* in Sitka firs was effectively controlled by thorough spraying (Johnson 1965) with Thiodan EC (1,000 g a.i./100 L of water).

Sounders and Barstow (1970) applied Thiodan EC at about 500 g a.i./ha with excellent results against the gall aphid *Adelges cooleyi.* Saunders and Barstow (1972) found Thiodan EC applied at a rate of 500 g a.i./ha by the ULV method to be highly effective against the aphid *Schizolachnus pineti.* Brown (1970) controlled the pine aphid *Pineus* sp. in East Africa by spraying with Thiodan (100 g a.i./100 L of water) plus Teepol. As a result of his tests Odera (1971) recommended the low-volume (atomizing) application of Thiodan EC (1,000 g a.i./100 L of water) with Teepol. This method controls all developmental stages of the pest. It is worth noting that as little as 500 g a.i./100 L of water applied as a spray gave 100% control of the eggs of this aphid.

Contarinia oregonensis attacks the seeds of Douglas firs. Johnson (1963) applied Thiodan EC at 500 g a.i./100 L of water by the low-volume technique after the cones opened and recorded good results in 2 yr of trials. Johnson (1964) confirmed the efficacy of Thiodan at a later date when he found that the seed yield of the treated plots was 2½ times as great as that of the untreated plots.

Goesswald (1958) mentioned the very strong action of Thiodan against termites such as *Calotermes flavicollis.*

Evidence of the efficacy of Thiodan EC at 62 g a.i./100 L of water plus oil against the *Eriophyidae* mite *Trisetacus campnodus* was given by Saunders and Barstow (1972).

Many other forest pests, including beetles, caterpillars, sawflies, and aphids can be controlled with Thiodan but these cannot be dealt with here.

f) Other uses of Thiodan: Control of tsetse fly

Vast areas of Africa are affected by the protozoan parasitic disease called trypanosomiasis. It is transmitted by the tsetse fly through its blood feeding activities. Man and domestic livestock are severely affected by different species of *Trypanosoma*—to the extent that an estimated 7 million km^2 at present denied development could be opened up for meat production.

In the early 1960s, Burnett (1963), through topical studies showed the tsetse fly's exceptional susceptibility to Thiodan. Since these early days Thiodan has become the leading product in the aerial application against tsetse but in two different techniques.

One technique classically uses helicopters to apply a single high residual dose of Thiodan to dry season tsetse habitats (Spielberger *et al.* 1977). Thiodan 25 ULV is applied at rates of about 800 to 1,000 g a.i./ha using electrically driven atomizers producing droplets of a VMD of 150 microns. Only 10 to 20% of the total area is sprayed and the method has been used chiefly in Nigeria and Cameroon.

The other technique relies on sequential low-dosage aerosol applications of Thiodan classically from fixed-wing aircraft. Four to 6 applications of nonresidual aerosol of 30 microns VMD are made at 10 to 20 day intervals depending on temperature. Swath intervals are 200 to 400 m and the aerosol released under inversion condition drifts a long way through the woodland causing considerable overlap. The kill is thought to be effected chiefly by direct airborne impact on to the flies.

Thiodan ULV is used at a dosage of normally 10 to 20 g a.i./ha. These extremely low dosages and the high susceptibility of tsetse to Thiodan make the technique safe environmentally (Cockbill 1979, Magadza 1978). It has also other benefits since large areas can be treated with great efficiency in a short time and at relatively low cost. The first large scale operation was in Zambia in 1968 (Park *et al.* 1972) but in the last decade, the technique has also become part of control programs in Botswana (Davies and Bowles 1979), Zimbabwe-Rhodesia (Chapman 1976), Nigeria, Tanzania, and the Ivory Coast.

List of pests mentioned in this publication

Acanthomia sp.
Acaphylla theae
Aceria mangiferae
Aceria sheldoni
Achaea janata

Acrosternum hilare
Acyrtosiphon pisum
Acyrtosiphon sp.
Adelges cooleyi
Aelia rostrata

Agriotes sp.
Agromyza atricornis
Agrotis sp.
Alabama argillacea
Aleurocanthus woglumi
Amsacta moorei
Amsacta sp.
Anarsia ephippias
Anarsia lineatella
Anomis sabulifera
Antestiopsis sp.
Antestiopsis lineaticollis
Anthonomus eugenii
Anthonomus grandis
Anthonomus pomorum
Antigastra catalaunalis
Anticarsia gemmatalis
Anuraphis roseus
Aonidiella aurantii
Aphidula schneideri
Aphis craccivora
Aphis fabae
Aphis gossypii
Aphis pomi
Aphytis africanus
Apion sp.
Apion corchori
Apion flavipes
Apsylla cistellata
Ascotis sp.
Ascotis selenaria
Athalia lugens

Balaninus nucum
Baris deplanata
Bathycoelia sp.
Batrachedra amydraula
Bemisia tabaci
Biston suppressarius
Bixadus sierricola
Buprestis sp.
Busseola fusca
Brevicoryne brassicae
Bruchus rufimanus
Calacarus carinatus
Calotermes flavicollis
Cecidophyopsis (Phytoptus) ribis
Cemiostoma scitella

Cephonodes hylas
Cerambycidae
Ceratocystis fimbriata
Ceutorhynchus assimilis
Chalcodermus aeneus
Cheimatobia brumata
Chilo sp.
Chilo auricilia
Chilo tumidicostalis
Chilo zonellus
Chilotraea sp.
Chilotraea infuscatella
Chromaphis juglandicola
Cirphis sp.
Cirphis unipuncta
Cleonus mendicus, see:
       Temnorhinus mendicus
Clysia ambiguella
Cnaphalocrocis medinalis
Colias lesbia
Contarinia mori, see:
       Diplosis mori
Contarinia oregonensis
Contarinia sp.
Cryptomyzus ribis
Cyrtopeltis tenuis

Dactynotus carthami
Darna trima
Dasychira securis
Dasyneura brassicae
Dasyneura lini
Dasyneura tetensi
Dasynus piperis
Dendroctonus pseudotsugae
Diacrisia sp.
Diacrisia obliqua
Dialeurodes citri
Diaphorina citri
Diatraea sp.
Diatraea saccharalis
Dichocrocis sp.
Diconocoris hewitti
Diparopsis sp.
Diparopsis watersi
Diplogomphus hewitti, see:
       Diconocoris hewitti
Diplosis mori

Distantiella theobroma
Drosicha mangiferae

Earias sp.
Earias biplaga
Earias insulana
Empoasca sp.
Empoasca fabae
Epicampoptera sp.
Epicauta sp.
Epilachna varivestis
Epitrimerus vitis
Epitrix cucumeris
Erannis defoliaria
Eriophyes essigi
Eriophyes (Phytoptus) vitis
Eriophyidae
Eriosoma lanigerum
Etiella zinckenella
Euschistus heros
Eugnamptus marginatus
Euproctis sp.
Euproctis chrysorrhoea
Euproctis lunata
Eurystylus capensis
Euxoa sp.
Euxoa auxiliaris

Frankliniella fusca

Glossina sp.
Glossina morsitans
Glossina pallidipes
Glossina swynnertoni
Gnathotrichus materarius
Gnathotrichus sulcatus

Habrochila ghesquierei
Haltica sp.
Haltica ampelophaga
Heliothis sp.
Heliothis armigera
Heliothis virescens
Helopeltis sp.
Helopeltis theivora
Hemitarsonemus latus
Hispa armigera
Holotrichia impressa

Homoeosoma electellum
Hyalopterus pruni
Hylemya coarctata
Hylobius pales
Hylurgops pinifex
Hypera postica
Hypothenemus hampei
Hyponomeuta malinella

Idiocerus atkinsoni
Idiocerus clypealis
Idiocerus niveosparsus
Ips calligraphus
Ips grandicollis
Ips pini
Leptinotarsa decemlineata
Leptispa pygmaea
Leptocorisa sp.
Leucoptera scitella, see:
    Cemiostoma scitella
Liothrips oleae
Lipaphis erysimi
Lithocolletis blancardella
Longitarsus sp.
Lophobaris pipcris
Loxostege commixtalis
Loxostege sticticalis
Lygus sp.
Lymantria dispar
Macrosiphum euphorbiae
Macrosiphum sonchi
Malacosoma neustria
Mamestra brassicae
Manduca sexta
Melanagromyza phaseoli
Melanitis leda ismene
Meligethes aeneus
Melolontha sp.
Microtermes sp.
Microtermes heimi
Monochamus sp.
Musca domestica
Myllocerus sp.
Myllocerus maculatus
Myzus persicae

Nephotettix sp.
Nezara viridula

Numicia viridis
Nymphula depunctalis
Nysius vinitor

Oberea brevis
Odontotermes sp.
Oecophylla smaragdina
Oligonychus mangiferus
Omnatissus binotatus
Oregma lanigera
Orthops sp.
Orthotomicus caelatus
Oscinella frit
Ostrinia nubilalis
Panonychus citri
Papilio sp.
Parasa sp.
Pectinophora gossypiella
Pegomya betae
Pelopidas agna
Perigea capensis
Peregrinus maidis
Perileucoptera coffeella
Phoridae
Phthorimaea operculella
Phyllobius oblongus
Phyllocoptes vitis
Phyllotreta sp.
Phytonomus variabilis, see:
        Hypera postica
Phytoptus avellanae
Phytoptus ribis, see:
        Cecidophyopsis ribis
Phytoptus vitis, see:
        Eriophyes vitis
Pieris sp.
Pieris brassicae
Pieris rapae
Pineus sp.
Pissodes approximatus
Pissodes sitchensis
Pityogenes hopkinsi
Platyedra gossypiella, see:
        Pectinophora
Plusia sp.
Plusia acuta
Plusia arichalcea
Plutella maculipennis

Polychrosis botrana
Prays citri
Prays oleellus
Proceras indicus
Protoparce sexta, see:
        Manduca sexta
Pseudaletia separata
Pseudoplusia includens
Psylla sp.
Psylla mali
Psylla pyricola
Pyrilla perpusilla
Pugria sp.

Rhopalosiphum maidis
Rhopalosiphum pseudobrassicae

Sahlbergella singularis
Sanninoidea exitiosa
Sappaphis mali
Schizaphis graminum
Schizolachnus pineti
Schizura concinna
Sciara sp.
Scirpophaga nivella
Scirtothrips sp.
Scirtothrips aurantii
Scirtothrips dorsalis
Selenocephalus virescens
Sesamia sp.
Sesamia inferens
Setora nitens
Sitona lineata
Sogatella sp.
Sparganothis pilleriana
Spodoptera sp.
Spodoptera exempta
Spodoptera exigua
Spodoptera littoralis
Spodoptera litura, see:
        Spodoptera littoralis
Steneotarsonemus ananas
Stephanoderes hampei, see:
        Hypothenemus hampei
Syllepta derogata
Synanthedon sp.
Synanthedon hector
Synanthedon pictipes

Tarsonemidae
Tarsonemus pallidus
Temnorhinus mendicus
Tetranychidae
Tetranychus cinnabarinus
Therioaphis maculata
Thosea sp.
Thosea bisura
Thrips sp.
Thrips tabaci
Tirathaba mundella
Tomicus piniperda
Toxoptera sp.
Toxoptera aurantii

Toxoptera citricida
Trialeurodes vaporariorum
Trichoplusia ni
Trioza erythreae
Trisetacus campnodus
Trypanosomes
Tryporyza sp.
Tychius flavus
Typhlocyba nigrobilineata

Vasates lycopersici
Vasates schlechtendali

Xyleborus sp.
Xyleborus ferrugineus

References

Abbott, D. C., D. C. Holmes, and J. O' G. Tatton: Pesticides residues in the total diet in England and Wales, 1966-1967: II-Organochlorine pesticide residues in the total diet. J. Sci. Food Agr. **20**, 245 (1969).

Abu Yaman, I. K.: Chemical control of the alfalfa weevil, *Hypera postica* Gylh., in Saudi Arabia. Unpublished report (1970).

Ahmad, F., and M. D. Mahsin: Control of cotton bollworm *Heliothis armigera* (Hb.) by air in Multan district of West Pakistan. Internat. Pest Control 11, 14 (1969).

Alam, M. Z.: Insect pests of rice in East Pakistan and their control. Agriculture Information Service, Dacca-3, p. 99. East Pakistan (1965).

Al-Azawi, A. F.: Studies on the effect of *Bacillus thuringiensis* Berl. on the spiny bollworm, *Earias insulana* Boisd. and other lepidopterous insects. Entomophaga **9**, 137 (1964).

Ali, S.: Degradation and environmental fate of endosulfan isomeres and endosulfan sulfate in mouse, insect and laboratory model ecosystem. Diss. Abstr. Int. B. **39**, 2117 (1978).

Allen, D. C., and J. A. Rudinski: Effectiveness of Thiodan, Sevin, and Lindane on insects attacking freshly cut douglas-fir logs. J. Econ. Entomol. **52**, 482 (1959).

Allen, W. W., H. Nakakihara, and G. A. Schaeffers: The effectiveness of various pesticides against the cyclamen mite on strawberries. J. Econ. Entomol. **50**, 648 (1957).

Almeida, P. R., and R. D. Cavalcante: Field test with new organic insecticides in the control of the coffee berry borer—*Hypothenemus hampei* (Ferr. 1867). Arquivos do Instituto Biológico, São Paulo 31 (3), 85 (1964).

――――― ―――――, and A. A. Holanda: Control do Acaro Branco do Algodoeiro com diversos produtos em polvilhamento e pulverização. Anais da X Reuniao de Fitosanitaristas do Brazil, Jan. (1966).

Anderson, L. D., and H. T. Reynolds: A comparison of the toxicity of insecticides for the control of earworm on sweet corn. J. Econ. Entomol. **53**, 22 (1960).

Ando, K.: Control of cherry tree borer, *Synanthedon hector* Butler, (Aegeridae: Lepidoptera), by branch and trunk spray of endosulfan. Proc. Kansai Plant Prot. Soc., No. 10 (1968).

Angelini, A., and P. Vandamme: Onze années d'experimentation insecticide en culture cotonnière de Côte d'Ivoire. Coton. Fibres Trop. 2, 531 (1965).

Anonymous 1: Tropical Pesticides Research Inst., Annual Report p. 45 (1964).

_____ 2: Proceedings of the 4th National Conference on Agriculture and Biology Plant and Biological Science, and Animal Science Sect. Kasetsart Univ., Banghen, Bangkok, 27-29 Jan. (1965).

_____ 3: The recent outbreak of armyworm *Spodoptera exempta* Wlk. in Northern Tanzania. Aug. 1965. Agr. Div. N. Research Centre, Tengeru, Arusha/Tanzania. Tengeru Report No. 59 (1965).

_____ 4: Chemicals for effective control of wooly aphis. Exp. Station Bull., Philippines, Nov.-Dec. (1965).

_____ 5: The life history and control of padi stem borers in West Malaysia. Entomology Division Research Branch, Div. Agr. Kuala Lumpur, Tech. Leaflet No. 1, Dec. (1967).

_____ 6: Agricultural Bank of Sudan, Khartoum. Thiodan Recommendation, Letter 12/3/67. Unpublished (1967).

_____ 7: Agricultural Bank of Sudan, Khartoum. Directive on spraying of cotton with pesticide in schemes financed by the Bank for Season 67/68. Unpublished (1968).

_____ 8: A jowar with a promise. Intensive Agr. 7, 26 (1969).

_____ 9: Chemical control of sorghum pests. Extract from Entomology Sect.: Proceedings of the All India Workshop on Sorghum and Millets, Hyderabad (Apr. 28-May 3, 1970) and Progress Reports of the All India Sorghum and Millets Improvement Projects, p. 107 (1969-70).

_____ 10: Régéneration d'une proche à Mirides. Hoechst Afrique de l'Quest "Cameroun", Douala, 1968 (Institut Francais du Café et du Cacao (IFCC). Unpublished report (1970).

_____ 11: Mise au point d'une méthode traitement antimirides par ULV. IFCC, Rapport d'Activité 1971, Côte d'Ivoire, Annual report, Institut Francais du Café et du Cacao, 40/41 (1971).

_____ 12: Ensayos contra Paulilla en el Trigo. Hoechst Iberica S.A., Delegación de Sevilla. Unpublished report (1971).

_____ 13: Aplicaciones Thiodan 3 espolvoreable. Ministerio de Agricultura, Dirección General de Agricultura, Sección de Protección de Cultivos, 3.5. (Official Approval) (1972).

_____ 14: Effectivity of different insecticides for control of *Diplosis mori* Yokoyama (mulberry shoot gall midge) on *Morus alba* L (mulberry), Korea 1971 and Effectivity of different insecticides for control of *Baris deplanata* Roelofs (mulberry small weevil) on *Morus alba* L. (mulberry), Korea. Unpublished reports (1971).

_____ 15: Aplicaciones Thiodan 35 emulsionable, Ministerio de Agricultura, Madrid, 3.5. (Official approval) (1971).

_____ 16: Registration Thiodan dust 3, No. de registro 9869/77 Ministerio de Agricultura, Madrid. (Official approval) (1972).

_____ 17: Aplicaciones de Thiodan. Ministerio de Agricultura, Madrid 15.9. (Official approval) (1972).

_____ 18: Revisión de la Literatura sobre el Control Quimico de la Broca, La Broca del Fruto del Café, Asociación Nacional del Café, Boletin No. 11, Guatemala, Oct., p. 56 (1972).

_____ 19: 1972 Pest and Disease Control Program for Cole Crops. Univ. Calif., Div. Agr. Sci. (1972).

_____ 20: 1972 Pest and Disease Control Program for Oil Seed Crops. Univ. Calif., Div. Agr. Sci. (1972).

_____ 21: Halt the pest on gram today. Intensive Agr. 10, 19 (1972).

_____ 22: Protokoll über die Durchführung eines Versuches (Testes) Prüfung Nr. 37/1972. Zentrales Kontroll- und Prüfungsinstitut für Landwirtschaft in Prag., Abt. für Quarantäne und Pflanzenschutz in Brno., Zemedelska 1a. Unpublished report (1972).

_____ 23: 1973 Pest and Disease Control Program for Walnuts. Univ. Calif., Div. Agr. Sci. (1973).

_____ 24: Emprege de diferentes insecticidas no controle da "Broca" do Café, Regioes do Estado de Minas Gerais. "Resumos" of First Brazilian Congress about insects and diseases of coffee, Vitoria, Brazil, Julio 4-6, p. 82 (1973).

_____ 25: Sonderdruck Württembergisches Wochenblatt für Landwirtschaft, aus Nr. 12 vom 21.3. (Stuttgart) (1970).

Archer, T. E.: Endosulfan residues on alfalfa hay exposed to drying by sunlight, ultraviolet light and air. Pest. Sci. 4, 59 (1973).

_____, Nazer, J. K., and D. G. Crosby: Photodecomposition of endosulfan and related products in thin films by ultraviolet light irradiation. J. Agr. Food Chem. 20, 954 (1972).

Aronica, C. S.: Sam Sam demonstration of ULV aerial application of Thimul 35 ULV for the control of cotton pests in 1970/71 season, Sudan. Unpublished report (1970/71).

Atkins, E. L., E. A. Greywood, and R. L. MacDonald: Toxicity of pesticides and other agricultural chemicals to honey bees. Univ. Calif. Rev. 9/73 (1973).

Awasthi, M. D., K. N. Karla, and R. S. Dewan: Residues of insecticides in sugarcane juice and gur following chemical protection of crop from insects. Pesticides 11, 33 (1977).

Baggiolini, M., E. Guinard, E. A. Carlen, C. Desbaillet, and P. Kristof: Gegenwärtige Möglichkeiten zur Bekämpfung der Kräuselmilbe im Weinbau. Schweiz. Z. Obst u. Weinbau 107, 359 (1971).

_____ _____ _____ _____ _____ Possibilitées actuelles de lutte contre l'acariose de le vigne (*Phytoptus vitis* Nd.). Rev. suisse viti. arbori. 2, 9 (1970).

Bakhetia, D. R. C., and A. S. Sidhu: Studies on the chemical control of the bihar hairy caterpillar, *Diacrisia obliqua* Walter (Arctiidae: Lepidoptera). Fourth All Indian Workshop on Pulse Crops, April, p. 4 (1970).

Ballschmiter, K., und G. Tölg: Metabolismus des Thiodans in Insekten. Angew. Chem. 78, 775 (1966).

Barbieri, F.: Impiego del Thiodan per la Lotta contre L'Eriofide del Nocciuolo. Atti de "La Giornata della Difesa Fitosanitaria del Nocciuolo in Compania", Lauro (Avellino) giugno, p. 46 (1966).

Barnes, W. W., and G. W. Ware: The absorption and metabolism of C^{14}-labeled endosulfan in the house fly. J. Econ. Entomol. 58, 286 (1965).

Barth, E.: Zwei wichtige Schädlinge der Johannisbeeren. (Pflanzenschutzdienst Baden-Württemberg Warndienstmitteilung) Württ. Wochenbl. Landw. 137, Nr. 12 (1970).

Bauer, K.: Studium über Nebenwirkungen von Pflanzenschutzmitteln auf Fische und Fischnährtiere. Mitt. Biol. Bundesanst. Land-Forstwirtsch. (Berlin-Dahlem) Heft 105 (1961).

Bauer, U.: Das Verhalten von einigen Herbiziden und Insektiziden bei der Aufbereitung von Oberflächenwasser zu Trinkwasser. Vom Wasser 39, 161 (1972).

_____ Anreicherung von insektiziden Chlorkohlenwasserstoffen und PCB in Algen. Schriftenr. Ver. Wasser-, Boden- Lufthyg., Berlin-Dahlem, H. 37, 211 (1972).

Beard, J. E., and G. W. Ware: Fate of endosulfan on plants and glass. J. Agr. Food Chem. 17, 216 (1969).

Beck, E. W., J. C. Johnson Jr., D. B. Woodham, D. B. Leuck, L. H. Dawsey, J. E. Robbins, and M. C. Bowman: Residues of endosulfan in meat and milk of cattle fed treated forages. J. Econ. Entomol. 59, 1444 (1966).

Becker, G.: Gaschromatographische Simultanbestimmung von chlorierten Kohlenwasserstoffen und Phosphorsäureestern in pflanzlichem Material. Deutsche Lebensmittel-Rundschau 67, 125 (1971).

Becker, W. B.: Tests with endosulfan to prevent borer damage to unseasoned pine logs. J. Econ. Entomol. 57, 166 (1964).

Belliveau, P. E., V. Mallet, and R. W. Frei: A spray method for the fluorescent detection of sulphur-containing organic compounds. J. Chromatog. 48, 478 (1970).

Beran, F.: Der gegenwärtige Stand unserer Kenntnisse über die Bienengiftigkeit und Bienengefährlichkeit unserer Pflanzenschutzmittel. Gesunde Pflanze 22, 21 (1970).

Bertuzzi, P. F., J. Kamps, and C. I. Miles: Extraction of chlorinated pesticide residues from nonfatty samples of low moisture content. J. Assoc. Off. Anal. Chemists 50, 623 (1967).

Bhattacherjee, N. S., S. K. Handa, and A. K. Dikshit: Efficacy of insecticides on soybean crop. Entomol. Newsletter 4, No. 1 (1974).

Bichoo, S. L.: Insect pests and diseases of soybean and their control. Farmer and Parliament 5, 13 (1970).

Bliss, M., jr., and W. H. Kearby: Evaluation of Dieldrin, dimethoate and endosulfan as stump sprays for control of the pales weevil and northern pine weevil in Central Pennsylvania. J. Econ. Entomol. 63, 341 (1970).

Bluman, N.: Health protection branch department of national health and welfare, Vancouver, B.C. Personal communication related to K. A. McCully, Health Prot. Br. Dept. Nat. Health and Welfare, Ottava, Ontario (1973).

Boncato, A. A., and I. M. Gandia: Effect of four spraying frequencies with six insecticides in the control of the coffee berry borer (*Stephanoderes hampei* Ferr.) Philipp. J. Plant Ind. **33**, 109 (1968).

Bonnemaison, L.: Essais Insecticides de Laboratoire sur les Chenilles et les Oeufs d'*Adoxophytes reticulana* Hb. (Tortricidae) et de *Mamestra brassicae* L. (Noctuidae). Phytiatr.-Phytopharm. **21**, 171 (1972).

Bot, J., N. d. L. Genis, and N. Hollings: A guide to the use of pesticides and fungicides in South Africa. Insecticide Res. Sect., Plant Prot. Res. Inst., Pretoria (1973).

Bowman, M. C., and M. Beroza: A copper-sensitized, flame-photometric detector for gas chromatography of halogen compounds. J. Chromatogr. **7**, 484 (1969).

_____ _____ Use of Dexsil 300 on a specially washed Chromosorb W for multicomponent residue determination of phosphorus- and sulfur containing pesticides by flame photometric GLC. J. Assoc. Off. Anal. Chemists **54**, 1086 (1971).

Brace, N. O.: Preparation and Diels-Alder reactions of 2,5-dihydrofuran. J. Amer. Chem. Soc. **77**, 4157 (1955).

Brader, L.: L'Efficacité de quelques insecticides vis-à-vis de chenilles de la capsule, *Diparopsis watersi* (Roths.) et *Heliothis armigera* (Hb.). Coton Fibres Tropic. **23**, 483 (1968).

Bressau, G.: Beurteilung von Pestizidrückständen in Fleisch und Fleischerzeugnissen. Deutsche Forschungsgemeinschaft, Forschungsbericht "Rüchstände in Fleisch und Fleischerzeugnissen", 84 pp. (1975).

Bro-Rasmussen, F., F. Rodin, und K. Voldum-Clausen: Überprüfung einer gaschromatographischen Methode zur Bestimmung chlorierter Insektizide in Butter und pflanzlichen Erzeugnissen. Z. Lebens.-Unters.-Forsch. **138**, 276 (1960).

Brown, K. W.: Tests of insecticides against *Pineus* sp. in East Africa. East Afr. Agr. For. J. **36**, 200 (1971).

Buhl, C., und M. Waede: Bericht über einen Thiodan-Kaltnebeleinsatz vom Hubschrauber aus zur Bekämpfung der Kohlschotenmücke (*Dasyneura brassicae* Winn.) und des Kohlschotenrüsslers *Ceutorhynchus assimilis* Payk.) in blühenden Rapsbeständen. Nachrichtenbl. Deutsch. Pflanzenschutzdienstes (Braunschweig) **14**, 38 (1962).

_____, and F. Schütte: Über das Auftreten und Bekämpfung von *Apion flavipes* Payk. in Weissklee-Samenbeständen. Z. Pflanzenkr. Pflanzenschutz **71**, 189 (1964).

Burke, J. A., and L. M. Porter: Note on the effect of sample moisture content on extraction of TDE from kale. J. Assoc. Off. Anal. Chemists **50**, 1260 (1967).

_____ _____, and S. J. V. Young: Evaluation of two extraction procedures for pesticide residues resulting from foliar application and root absorption. J. Assoc. Off. Anal. Chemists **54**, 42 (1971).

Burnett, G. F.: Topical applications of insecticides to tsetse flies. VIII-possible correlations between toxicities to *Musca domestica* and *Glossina morsitans*. Tropical Pesticides Res. Inst., Arusha, Tanganyka Misc. Rep. No. 357, p. 7 (1962 a).

_____ Topical applications of insecticides to tsetse flies. VII–Thiodan and some more organophosphates to young *G. morsitans.* Tropical Pesticides Res. Inst., Arusha, Tanganyka, Misc. Rep. No. 342, p. 3 (1962 b).

_____ The susceptibility of tsetse flies to topical applications of insecticides. VI–Data on more chlorinated hydrocarbon- and organophosphates, and a general discussion. Bull. Entomol. Res. 53, 753 (1963).

Byrn, S. R., and P. Y. Siew: Structure and confirmation of β-Thiodan in the solid state and in solution. Application of the infrared–X-ray method. J. Chem. Soc. Perkin Trans. 2, 144 (1977).

Caballero Ga. de Vinuesa, J. I.: Prueba comparativa contra prays de olovo. Ministerio de Agricultura, Delegación Provindial de Sevilla, Sección Agronómica. Unpublished report (1971).

Calora, F. B., and M. P. Ferrino: Some chlorinated, phosphatic and carbamate insecticides in the control of rice stemborers. Coll. of Agr. and Central Expt. Sta., Univ. Philippines 48, No. 1, June (1964).

Carey, A. E.: Monitoring pesticides in agricultural and urban soils of United States. Pest. Monit. J. 13, 23 (1979).

Carlson, E. C.: New insecticides to control sunflower moth. J. Econ. Entomol. 64, 208 (1971).

Carnegie, A. J. M.: Report on a visit to Swaziland and a brief review of a present status of *Numicia viridis* Muir. Mount Edgecombe, Exp. Sta. S. Afr. Sugar Assoc. 18 (1967).

Cassil, C. C., and P. E. Drummond: A plant surface oxidation product of endosulfan. J. Econ. Entomol. 58, 356 (1965).

Celino, C. S., and U. V. Molino: Studies on the field control of *Diaphorina citri,* Kuway. Animal Husb. Agr. J. 6, 23 (1971).

Chadwick, P. R., J. S. S. Beesley, P. J. White, and H. T. Matechi: An experiment on the eradication of *Glossina swynnertoni* Aust. by insecticidal treatment of the resting sites. Tropical Pest. Res. Inst. Misc. Rep. No. 443 (1964).

Chan Check Onn: Some notes on the oil palm bunch moth *Tirathaba mundella* Walk. (Lepidoptera:Pyralididae). Reprint from International Oil Palm Conference, p. 5 (1972).

Chapman, N. G.: Aerial spraying of tsetse flies (*Glossina* sp.) in Rhodesia with Ultra Low Volumes of endosulfan. Trans. Rhodesian Sci. Assoc. 57, 12 (1976).

Chari, M. S., S. N. Seshadri, and H. K. Patel: Control of mangohoppers. Pesticides, Dec., p. 3 (1969).

Chatterjee, D. K.: Effect of few moderate to slightly toxic insecticides on some rice pest and on crop yield. Pesticides 6, 19 (1972).

Chau, A. S. Y.: Derivative formation for the confirmation of endosulfan by gas chromatography. J. Assoc. Off. Anal. Chemists 52, 1240 (1969).

_____ Confirmation of pesticide residue identity. I. Derivative formation for the confirmation of photoproducts of Endrin: Hexachloro- and pentachloro-ketone pesticide residues by gas chromatography. J. Assoc. Off. Anal. Chemists 55, 519 (1972).

_____, and M. Lanouette: Confirmation of pesticide residue identity. II Derivative formation in soil matrix for the confirmation of DDT, DDD and

Methoxychlor, Perthane, *cis-* and *trans-*Chlordane, Heptachlor and Heptachlor epoxide pesticide residues by gas chromatography. J. Assoc. Off. Anal. Chemists 55, 1058 (1972).

———, and K. Terry: Confirmation of pesticide residue identity. IV. Derivative formation in solid matrix for the confirmation of α- and β-endosulfan by gas chromatography. J. Assoc. Off. Anal. Chemists 55, 1228 (1972).

——— ——— Confirmation of pesticide residue identity. VI. Derivative formation in solid matrix for confirmation of heptachlor and endosulfan isomers. J. Assoc. Off. Anal. Chemists 57, 394 (1974).

———, and R. J. Wilkinson: Some separation characteristics of an OV-101/OV-210 column for organochlorinated pesticides with particular reference to the separation of photoendrin and endrin. Bull. Environ. Contam. Toxicol. 7, 93 (1972).

Cheng, H. H., and H. E. Braun: Chlorpyrifos, carbaryl, endosulfan, leptophos and trichlorfon residues on cured tobacco leaves from field-treated tobacco in Ontario. Can. J. Plant. Sci. 57, 689 (1977).

Chopra, N. M., and A. M. Mahfouz: Metabolism of endosulfan I, endosulfan II and endosulfan sulfate in tobacco leaf. J. Agr. Food Chem. 25, 32 (1977).

CIPAC: Analysis of technical and formulated pesticides. Handbook, Vol. 1 (1970).

Clinch, P. G., and D. U. Higgons: Thiodan—A new control for big bud mite. The Grower, Mar. 11, p. 484 (1961).

Cockbill, G. F.: Tests on the relative toxicity of certain commercial insecticides to tsetse fly. Unpublished report (1961).

——— The effect of ultra-low-volume aerial applications of endosulfan applied against *Glossina* (Diptera:Glossinidae) on populations of non-target organisms in Savanna woodland in Zimbabwe-Rhodesia. Bull. Entomol. Res. 69, 645 (1979).

Congressional Record-Senate: Washington, D.C., U.S.A., p. 4415, May 1 (1969).

Corneliussen, P. E.: Pesticide residues in total diet samples (IV). Pest. Monit. J. 2, 140 (1969).

——— Pesticide residues in total diet samples (V). Pest. Monit. J. 4, 89 (1970).

——— Pesticide residues in total diet samples (VI). Pest. Monit. J. 5, 313 (1972).

Crockett, A. B., G. B. Wiersma, H. Tai, W. G. Mitchell, P. F. Sand, and A. E. Carey: Pesticides in soil. Pesticide residue levels in soils and crops. FY-70-National Soils Monitoring Program II. Pest. Monit. J. 8, 69 (1974).

Dalela, R. C., M. C. Bhatnagar, A. K. Tyagi, and S. R. Verma: Adenosinetriphosphatase activity in few tissues of a freshwater teleost, *Channa gachua*, following in vivo exposure to endosulfan. Toxicol. 11, 361 (1978).

Davies, J. C.: Insecticides for the control of the spread of groundnut rosette disease in Uganda. PANS 21, 1 (1975).

———, and J. Bowles: Effect of large scale aerial applications of endosulfan on tsetse fly, *Glossina morsistans centralis* Machado, in Botswana. Centre for Overseas Pest Research, Misc. Rep. No. 45 (1979).

———, and W. R. Ingram: Insecticides on cotton in Uganda. Comparison of DDT plus Toxaphene, Endosulfan and Trichlorphon. Emp. Cotton Grow. Rev. 42, 300 (1965).

Deema, P., E. Thompson, and G. W. Ware: Metabolism, storage and excretion of ^{14}C-Endosulfan in mouse. J. Econ. Entomol. **59**, 546 (1966).

Demeter, J., and A. Heyndrickx: Reserved-phase liquid chromatography of endosulfan and transformation products. Meded. Fac. Landbouwwet. Rijksuni. Gent. **41/2**, 1431 (1976).

Depew, L. J.: Evaluation of foliar and soil treatments for greenbug control of sorghum. J. Econ. Entomol. **64**, 169 (1971).

Deshmukh, S. N., and H. S. Adarsh: Comparative toxicity of some commercial formulations of common insecticides to caterpillars of the cabbage butterfly, *Pieris brassicae* L. J. Res. **8**, 339 (1971).

Deutsche Forschungsgemeinschaft: Methodensammlung zur Rückstandsanalytik von Pflanzenschutzmitteln Verlag Chemie, Weinheim/Bergstr. (1968/1972).

Dewitt, J. B., W. H. Stickel, and P. F. Stringer: Wildlife studies: Patuxent Wildlife Research Centre, U.S. Dept. Int., Fish and Wildlife Serv. Circ. **167**, 74 (1963).

Diehl, O.: Wirksame Bekämpfung des Kohlschotenrüsslers und der Kohlschotenmücke durch Thiodan-Kaltnebel in Ölfruchtbeständen. Gesunde Pflanz. **13**, 120 (1961).

Diemair, W., G. Maier, K. Pfeilsticker, und K. Schloegel: Beitrag zum Nachweis und zur Bestimmung von Pestiziden in pflanzlichen Lebensmitteln. II. Mitteilung Chromatographie pestizidhaltiger Pflanzenextrakte. Z. Lebensm.-Unters.-Forsch. **139**, 67 (1968).

Dieter, A.: Möglichkeiten zur sicheren Bekämpfung der Pocken- oder Gallmilbe (*Eriophyes vitis*) der Rebe. Landes Lehr- und Forschungsanstalt für Wein- und Gartenbau, Neustadt/Weinstr. Sonderdruck, p. 5 (1962-1965).

Dikshith, T. S. S., and K. K. Datta: Endosulfan: Lack of cytogenetic effects in male rats. Bull. Environ. Contam. Toxicol. **20**, 826 (1978).

Dizon, R. L., and A. J. Atienza: Experiment IV: Field evaluation of several insecticides for the control of insect pests of tomato. Unpublished report (1971).

Dochkova, B.: The cabbage moth on sugar beet and chemicals for its control. Rastit. Zash. (Bulgarien) **17**, 30 (1969).

Domanski, J. J., and F. E. Guthrie: Pesticide residues in 1972 cigars. Bull. Environ. Contam. Toxicol. **11**, 312 (1974).

Doom, D., and L. Luitjes: Bestrijding van de dennenscheerder (*Tomicus piniperda*) door stambespuitingen. Ned. Bosbouw-Tijdschr. **12**, 297 (1970).

Dorough, H. W., and J. R. Gibson: Chlorinated insecticide residues in cigarettes purchased 1970-72. Environ. Entomol. **1**, 739 (1972).

_____, K. Huhtanen, T. C. Marshall, and H. E. Bryant: Fate of endosulfan in rats and toxicological considerations of apolar metabolites. Pest. Biochem. Physiol. **8**, 241 (1978).

Douthwaite, R. J.: Vortrag gehalten auf dem 24. Kurs des British Council (Monitoring the side effects of pesticide use) 8.-19.04.80-Egham, U.K. (1980).

Duggan, R. E., and P. E. Corneliussen: Dietary intake of pesticide chemicals in the United States (III), June 1968-April 1970. Pest. Monit. J. **5**, 331 (1972).

Dustan, G. G.: Effect of endosulfan (Thiodan) on geese. Pest. Progress (Canada) **3**, 6 (1965).

Ebing, W.: Verfahren zur Erzielung gut reproduzierbarer R_f-Werte bei serienmäßig durchgeführter Dünnschichtchromatographie, Routinemethode zur Identifizierung insektizider Chlorkohlenwasserstoffe. J. Chromatogr. 44, 81 (1969).

_____ Gaschromatographie der Pflanzenschutzmittel. Tabellarische Literaturreferate III. Mitt. Biol. Bundesanst., Berlin-Dahlem, Heft 152 (1973).

_____ Gaschromatographie der Pflanzenschutzmittel. Tabellarische Literaturreferate IV. Mitt. Biol. Bundesanst., Berlin-Dahlem, Heft 161 (1974).

Eghtedar, E.: Die Empfindlichkeit von *Philonthus fuscipennis*, Mannh. und *Tachyporus hypnorum* L. (Col. Staphilinidae) gegenüber Insektiziden. Nachrichtenbl. Deutsch. Pflanzenschutzdienstes (Braunschweig) 21, 182 (1969).

Eichelberger, J. W., and J. J. Lichtenberg: Persistence of pesticides in river water. Environ. Sci. Technol. 5, 541 (1971).

Elkins, E. R., R. P. Farrow, and E. D. Kim: The effect of heat processing and storage on pesticide residues in spinach and apricots. J. Agr. Food Chem. 20, 286 (1972).

Elmore, C., and D. Magor: Insecticide experiments to control green peach aphid and pepper weevil on peppers. J. Econ. Entomol. 55, 375 (1962).

Elzner, J.: Zu der Analytik hydrophiler Metaboliten des Pflanzenschutzmittels Endosulfan und ihrer Ausscheidung bei Warmblütern (Ratten). Dissert. Johannes-Gutenberg-Universität, Fachbereich Chemie, Mainz (1973).

El Zorgani, G. A., and M. H. Omer: Metabolism of endosulfan isomers by *Aspergillus niger*. Agricultural Research Corporation, Wad Medani, Sudan. Internal report (1973).

EMBRAPA (Empresa Brasileira de pesquisa Agropecuária): Centro Nacional de Pesquisa de Soja. Comunicado tecnico No. 02, Set. (1979).

Emmel, L.: Die Wirkung von Thiodan auf die Blutlaus (*Eriosoma lanigerum* HAUSM.) und die Blutlauszehrwespe (*Aphelinus mali* HALD.). Anz. Schädlingsk. 31, 121 (1958).

Engel, H.: Erfolgreiche Maikäferbekämpfung. Badische Bauernzeitung 18, No. 24 (1965).

_____ Der Zünsler frisst unsere Ernten. Badische Bauernzeitung 24, 20 (1971).

_____ Erfolge und Probleme der Maiszünslerbekämpfung. Gesunde Pflanze 24, 47 (1972).

Epstein, E., and W. J. Grant: Chlorinated insecticides in run off water as affected by crop rotation. Soil Sci. Soc. Amer. Proc. 32, 423 (1968).

FAO/WHO: 1974 Evaluations of some pesticide residues in food. Report of the 1974 Joint Meeting of the FAO Working Party of Experts on Pesticide Residues and the WHO Expert Committee on Pesticide Residues. WHO Pest. Residues Series No. 4, 285 (1975).

Faubert Mounder, M. J. de, H. Egan, E. W. Godly, E. W. Hammond, J. Roburn, and J. Thompson: Clean up of animal fats and dairy products for the analysis of chlorinated pesticide residues. Analyst 89, 168 (1964).

Fehringer, N. V.: Effects of brine treatment on pesticide residues during recycling of pickling brine. J. Assoc. Offic. Anal. Chemists 61, 1441 (1978).

Feichtinger und Linden: DBP 1062252, Prior. 1958 (1958).

150   H. Goebel, S. Gorbach, W. Knauf, R. H. Rimpau, and H. Hüttenbach

Finkenbrink, W.: Über Thiodan, ein neues synthetisches Insektizid. Nachrichtenbl. Dtsch. Pflanzenschutzdienstes (Braunschweig) 8, 183 (1956).

Fonseca, E. J.: Trial reports Serere Research Station, P.O. Saroti/ Uganda. Min. Agr. and Co-Operatives, Dec. 17 (1971 a).

———— Pest attack on cotton at Serere 1971. Serere Experiment Committee Meeting 1971/72, Cotton Entomol. Unpublished report (1971 b).

Fonseca Ferrao, da A.: A broca dos frutos do café (Stephanoderes hampei Ferr.). Gazeta Agricola de Angola 5, No. 6 (1960).

Food Machinery Corporation, Niagara Chem. Div.: Unpublished laboratory reports (1964-1972).

Forman, S. E., B. L. Gilbert, G. S. Johnson, C. A. Erickson, and H. Adelman: Isotope-labeled insecticide Thiodan-5a, 9a-C_2^{14}. J. Agr. Food Chem. 8, 193 (1960).

Foschi, S., A. Cesari, J. Ponti, P. G. Bentirogly, and A. Bencivelli: Indagine sulla degradazione e morimento verticale dei fitofarmaci nel terreno. Not. Mal. Piante 82/83, 37 (1970).

Founk, J., and R. J. McClanahan: Laboratory studies on the toxicity of insecticides to larvae of the colorado potato beetle. J. Econ. Entomol. 63, 2006 (1970).

Frank, R., H. E. Braun, M. Holdrinet, D. P. Dodge, and S. J. Nepszy: Residues of organochlorine insecticides and polychlorinated biphenyls in fish from lakes Saint Clair and Erie, Canada 1968-76. Pest. Monit. J. 12, 69 (1978).

Frensch, H., und H. Goebel: DBP 1015797, Prior. 1954 DBP 960 989, Prior. 1954 (1954 a).

————, W. Finkenbrink, und W. Staudermann: DBP 963 282, Prior. 1954 (1954 b).

Gangaprasada Rao, N.: Sorghum culture, dry to irrigated farming. All-India Sorghum Improvement Project (1970) published by Reg. Res. Station (I.A.R.I.), Rajendranagar, Hyderabad (1970).

Geike, F.: Dünnschichtchromatographisch-Enzymatischer Nachweis und zum Wirkungsmechanismus von Chlorkohlenwasserstoff-Insektiziden. J. Chromatog. 44, 95 (1969).

———— Dünnschichtchromatographisch-Enzymatischer Nachweis und zum Wirkungsmechanismus von Chlorkohlenwasserstoff-Insektiziden. II. Nachweis durch Hemmung von Trypsin. J. Chromatog. 52, 447 (1970 a).

———— Insektizide- und antiesterase Wirkung von Chlorkohlenwasserstoff-Insektiziden nach UV-Bestrahlung. Z. angew. Entomol. 65, 98 (1970 b).

———— Dünnschichtchromatographisch-Enzymatischer Nachweis und zum Wirkungsmechanismus von Chlorkohlenwasserstoff-Insektiziden. III. Nachweis durch Phosphatase-Hemmung. J. Chromatog. 61, 279 (1971 a).

———— Dünnschichtchromatographisch-Enzymatischer Nachweis einiger Pestizide durch Hemmung der Amylase-Aktivität. J. Chromatog. 63, 343 (1971 b).

Gelosi, A., and P. Giunchi: Orientamenti de lotta contro il cleone della barbabietola (Temnorhinus mendicus Gyll.) Giornate Fitopatologiche (Udine, 11-14.5. 1971) (1971).

Gharib, A., and M. Djazayeri: Experimentations de 5 insecticides contre le *Batrachedra amydraula* Meyr. Entomol. Phytopathol. Appl. (Teheran) **29**, 36 (1969).

Gibson, J. R., G. A. Jones, A. W. Dorough, C. I. Lusk, and R. Thurston: Chlorinated insecticide residues in Kentucky burley tobacco, crop years 1963-1972. Pest. Monit. J. **7**, 205 (1974).

Goebel, H.: Über das Verhalten von Verbindungen der Endosulfanreihe bei UV-Bestrahlung in Wasser. Farbwerke Hoechst AG. Unpublished report (1971).

Goeke, G.: Gas-Chromatographie und Umweltschutz. Beitrag zur Gaschromatographischen Untersuchung von Wasser. Fette, Seifen, Anstrichm. **74**, 168 (1972).

Goesswald, K.: Zum Wirkungsmechanismus von Thiodan. Z. Angew. Zool. **45**, 129 (1958).

_____ Beitrag zur Wirkungsweise des Insektizids Thiodan. XI. Intern Kongr. Entomol., Wien 1960, Verhandlg. Bd. **II**, 605 (1962).

_____, E.-f. Schulze, und W. Kloft: Problems of application and action of Thiodan studied with S^{35}-labelled insecticide. Internat. Atomic Energy Agency Wien 1963, 241 pp. (1963).

Gorbach, S., und K. D. Bock: (Hoeschst AG), Unveröffentliche Reihenversuche (1974).

_____, und W. Knauf: Endosulfan und Umwelt. Das Rückstandsverhalten von Endosulfan in Wasser und seine Wirkung auf Organismen, die im Wasser leben. Schriftenr. Ver. Wasser-, Boden, -Lufthyg. (Berlin-Dahlem) Heft **34**, 85 pp. (1971).

_____, und E.-F. Schulze: Private Mitteilung (1974).

_____, R. Haaring, W. Knauf, und H. J. Werner: Residue analyses in the water system of East-Java. (River Brantas, Ponds, Sea-Water) after continued large-scale application of Thiodan in rice. Bull. Environ. Contam. Toxicol. **6**, 40 (1971 a).

_____ _____ Residue analyses and biotests in rice fields of East-Java treated with Thiodan. Bull. Environ. Contam. Toxicol. **6**, 193 (1971 b).

_____, O. E. Christ, H. M. Kellner, G. Kloss, und E. Boerner: Metabolism of endosulfan in milk sheep. J. Agr. Food Chem. **16**, 950 (1968).

Gould, H. J.: Alternatives to DDT for the control of blossom beetle on spring-sown oil-seed rape. Proc. 6th Brit. Insect. Fung. Conf. (1971).

Graham, J. R., J. Yaffe, T. E. Archer, and A. Bevenue: "Thiodan". In G. Zweig (ed.): Analytical methods for pesticides, plant growth regulators, and food additives, vol. **II**, p. 506. New York: Academic Press (1964).

Greve, P. A.: De Persistentie van Endosulfan in Oppervlaktewater. Meded. Fac. Landbouwwet. Rijksuniv. Gent **36**, 439 (1971).

_____, and S. L. Wit: Endosulfan in the Rhine river. J. Water Poll. Control Fed. **43**, 2338 (1971 a).

_____ _____ Rapid identification method for endosulfan from GLC peak shifts under the influence of alkali. J. Agr. Food Chem. **19**, 372 (1971 b).

Griffith, F., Jr., and R. V. Blanke: Microcoulometric determination of organo-chlorine pesticide in human blood. J. Assoc. Off. Anal. Chemists **57**, 595 (1974).

Guevara Calderon, J., and R. Moreno Dahme: Insecticidas recomendados para el combate de las plagas del algodonero en el ciclo agricola de 1972. Secretaria de Agricultura y Ganaderia, Mexico (1972).

Gupta, P. K.: Distribution of endosulfan in plasma and brain after repeated oral administration to rats. Toxicol. **9**, 371 (1978).

———, and R. C. Gupta: Pharmacology, toxicology and degradation of endosulfan. A review. Toxicol. **13**, 115 (1979).

———, and M. Ehrnebo: Pharmacokinetics of α- and β-isomers of racemic endosulfan following intravenous administration in rabbits. Drug. Metab. Dispos. **7** (1979).

———, K. M. Rai, and N. K. Joshi: Efficacy of some modern insecticides against wooly apple aphid, *Eriosoma lanigerum* (HSM). Indian J. Entomol. **31**, June, Part 2, p. 174 (1969).

Guthrie, F. E.: The nature and significance of pesticide residues on tobacco in tobacco smoke. Beitr. Tabakforsch. **4**, 229 (1968).

———, and T. G. Bowery: Thiodan and Telodrin residues on tobacco. J. Econ. Entomol. **55**, 1017 (1962).

——— ——— Pesticide residues on tobacco. Residue Reviews **19**, 31 (1967).

Hagen, A. F.: Evaluation of Thiodan and Sevin for control of webworm in sugar beets. J. Econ. Entomol. **54**, 799 (1961).

Harris, C. R., and W. W. Sans: Insecticide residues in soils on 16 farms in south-western Ontario—1964, 1966 and 1969. Pest. Monit. J. **5**, 259 (1971).

Harrison, R. B., D. C. Holmes, J. Roburn, and J. O' G. Tatton: The fate of some organochlorine pesticides on leaves. J. Sci. Food Agr. **18**, 10 (1967).

Hazleton Laboratories, Inc.: Thiodan ^{14}C-subacute feeding studies, dairy cows. Falls Church, Va., U.S.A. Unpublished report (1959).

Henderson, C. A., and F. M. Davis: Four insecticides tested in the field for control of *Diatraea grandiosella*. J. Econ. Entomol. **63**, 1459 (1970).

Hengy, H., and J. Thirion: The determination of Thiodan and Thiodan sulphate on tobacco and in smoke condensate. Beitr. Tabakforsch. **6**, 57 (1971).

Herbst, H. V.: Untersuchungen zur Toxizität von Endosulfan auf Fische und wirbellose Tiere. Schriftenr. Ver. Wasser-, Boden-, Lufthyg. Berlin-Dahlem, Heft **34**, 69 (1971).

Hernandez Paz, M.: Campana nacional para el control y possible erradicación de la plaga "Broca del Caféto". Rev. Cafétalera (Guatemala) 13-22, Julio (1972).

Herok, J.: Farbwerke Hoechst AG, unpublished internal report (1964).

Herzel, F.: Organochlorine insecticides in surface waters in Germany—1970 and 1971. Pest. Monit. J. **6**, 179 (1972).

Himel, Ch. M., and S. Uk: The dose-toxicity of chlorpyrifos and endosulfan insecticides on the house fly by topical, vapor and spray treatments as estimated by gas chromatography. J. Econ. Entomol. **65**, 990 (1972).

Hoch, P. E.: Tricyclic ketal compounds having biological activity. U.S. Pat. 3346596 (1967).

——— Bicyclic ketones having biological activity. U.S. Pat. 3 661 998 (1972).

Hocking, K. S., C. W. Lee, J. S. S. Beesley, and H. T. Matechi: Aircraft applications of insecticides in East Africa, XVI—Airspray experiment with endosulfan against *Glossina morsitans* Westw., *G. swynnertoni* Aust, and *G. pallidipes* Aust. Bull. Entomol. Res. (London) **56**, 737 (1966).

Hodgson, C. J.: Pests of citrus and their control. PANS **16**, 647 (1970).

Hoodless, R. A., J. A. Sidwell, J. C. Skinner, and R. D. Treble: Application of high-performance liquid chromatography to the determination of pesticides included in the European Economic Community Directive on fruit and vegetables. J. Chromatogr. **166**, 279 (1978).

Hoechst AG: Internal laboratory reports (1968-1972).

_____ Internal laboratory reports (1970-1973).

_____ Internal laboratory reports (1970-1975).

_____ Internal laboratory reports (1972).

_____ Internal laboratory reports (1980).

Hoechst Holland N.V.: Thiodan een belangrijk en interessant insekticide. Hoechst helpt **5**, Dec. (1967).

Hoffmann, J., und D. Eicheldoerfer: Zur Ozon-Einwirkung auf Pestizide der Chlorkohlenwasserstoffgruppe im Wasser. Vom Wasser **38**, 197 (1971).

Homonnay, F.: A nagyuzemi védekezés eredményer rajzó májusi cserebogarak ellen. Deutsche Übersetzung: Ergebnisse der grossbetrieblichen Vorbeugung gegen Maikäferschwärme. Die ungarische Landwirtschaft Frühjahr 1965, p. 11 (1965).

Hornig, H.: Erfahrungen eines Großeinsatzes zur Bekämpfung der Kohlschotenmücke (*Dasyneura brassicae* Winn.) und des Kohlschotenrüßlers (*Ceutorhynchus assimilis* Payk.) im Sprühverfahren vom Hubschrauber aus. Nachrichtenbl. Dtsch. Pflanzenschutzdienstes (Braunschweig) **14**, 40 (1962).

Houllier, M.: Tests de specialités insecticides pour le controle des mirides du cacaoyer. Unpublished report (1980).

_____ Données sur les possibilités d'utilisation du Thiodan pour le controle des mirides du cacaoyer. Unpublished report (1981).

Hoyt, C.: New materials for the control of apple rust mite. J. Econ. Entomol. **55**, 639 (1962).

Huettenbach, H.: Selective insecticides in integrated pest-control as illustrated by Thiodan (Endosulfan). Z. Pflanzenkr. Pflanzenschutz **76**, 667 (1969).

Huhtanen, K. L., and H. W. Dorough: Synthesis of ^{14}C-endosulfan from 2,3 ^{14}C-maleic anhydride. J. Labelled Comp. Radiopharm. **14**, 321 (1978).

Industrial Bio-Test Laboratories Inc., Northbrook, Ill. U.S.A.: Unpublished report of Dec. 30 (1965).

_____ Unpublished report of Dec. 5 (1967).

_____ Unpublished report of July 18 (1972).

Ingram, W. R.: An evaluation of several insecticides against berry borer and fruit fly in Uganda robusta coffee. East Afr. Agr. For. J. **30**, 259 (1965).

Innes, J. R. M., B. M. Ulland, M. G. Valerio, L. Petrucelli, L. Fischbein, E. R. Hart, A. J. Pallotta, R. R. Bates, H. L. Falk, J. J. Gart, M. Klein, I. Mitchell, and J. Peters: Bioassay of pesticides and industrial chemicals for tumorgenicity in mice. A preliminary note. J. Nat. Cancer Inst. **42**, 1101 (1969).

Ionescu, M., M. Tomoroga, and I. Tucra: Noi cercetari asupra bioecologiei si combaterii gargaritei lucernei (*Tychius flavus* Becker). (New investigations on the bio-ecology and control of the lucerne weevil (*T. flavus*). Anal. Inst. Cerc. Prot. 4, 271 (1966)

Johnson, N. E.: Insecticides tested for control of the douglas-fir cone midge. J. Econ. Entomol. 56, 236 (1963).

_____ Chemical control of the douglas-fir cone midge, *Contarinia oregon-ensis*, using a mistblower from a truck-mounted ladder. J. Econ. Entomol. 57, 556 (1964).

_____ A test of 12 insecticides for the control of the sitka-spruce weevil, *Pissodes sitchensis* Hopkins. J. Econ. Entomol. 58, 572 (1965).

Johnson, R. D., and D. D. Manske: Pesticide residues in total diet samples (IX). Pest. Monit. J. 9, 157 (1976).

Johnson, W. H., J. J. Domanski, T. J. Sheets, and Ch. S. Chang: Effects of freeze-drying on residues of TDE, DDT and Endosulfan in tobacco. J. Agr. Food Chem. 23, 117 (1975).

Jotwani, M. G., B. K. Rai, and S. Pradhan: Bio-assay of the comparative toxicity of some insecticides to the larvae of *Prodenia litura* Fabricius (Noctuidae: Lepidoptera). Indian J. Entomol. 23, 50 (1961).

Judge, F. D., and F. L. McEven: Field testing candidate insecticides on cole crops for control of cabbage looper and imported cabbageworm in New York State. J. Econ. Entomol. 63, 862 (1970).

Kadoum, A. M.: Modification of the micromethod of sample cleanup for thin-layer and gas chromatographic separation and determination of common organic pesticide residues. Bull. Environ. Contam. Toxicol. 3, 354 (1968).

_____ Partitioning method for sample cleanup for gas chromatographic analysis of common organic pesticide residues in biological materials. Bull. Environ. Contam. Toxicol. 4, 184 (1969).

Karla, A. N.: Efficiency of Endrin and Endosulfan granules in controlling top borer of sugarcane. Indian Sugar 20, No. 2 (1970).

Kaplan-Reiterer, O.: Versuch einer Maiszünslerbekämpfung über die Beregnung. Gesunde Pflanze 23, 28 (1971).

Karl, E.: Versuche zur chemischen Bekämpfung der Erdbeermilbe (*Tarsonemus pallidus* Banks). Nachrichtenbl. Dtsch. Pflanzenschutzdienst (Berlin) 18, 56 (1964).

Keil, J. E., C. B. Loadholt, B. L. Brown, S. H. Sandifer, and W. R. Sitterly: Decay of parathion and endosulfan residues on field-treated tobacco, South-Carolina 1971. Pest. Monit. J. 6, 73 (1972).

Keiser, J., and J. Tomikawa: Species-specific toxicity of certain insecticides to *Tephritids* in Hawaii suggested by unusual susceptibility relationships among oriental fruit flies, melon flies, and mediterranean fruit flies. J. Econ. Entomol. 63, 1756 (1970).

Keith, L. H., and H. L. Alford: Review of the application of nuclear magnetic resonance spectroscopy in pesticide analysis. J. Assoc. Offic. Anal. Chemists 53, 1018 (1970).

Keller, J. C., and T. T. Liang: The acute oral toxicities of some insecticides to american cockroaches. J. Econ. Entomol. 55, 144 (1962).

Khanna, R. N., D. Misra, M. Anand, and H. K. Sharma: Distribution of endosulfan in cat brain. Bull. Environ. Contam. Toxicol. 22, 72 (1979).

Klee, O.: Über den Einfluß der Temperatur und der Luftfeuchtigkeit auf die toxische Wirkung organisch synthetischer Insektizide. Z. angew. Zool. 47, 183 (1960).

Kloss, G., H.-M. Kellner, und O. Christ: Versuche an Schafen mit ^{14}C-markiertem Thiodan. Farbwerke Hoechst AG. Unpublished report (1966).

Knauf, W., und E.-F. Schulze: New findings on the toxicity of endosulfan and its metabolites to aquatic organisms. Meded. Fac. Lanbouwwet. Rijksuniv. Gent 38, 717 (1973 a).

_____ _____ Neue Ergebnisse zur Toxizität von Endosulfan und seinen Metaboliten bei Wasserorganismen. Farbwerke Hoechst AG. Unpublished report (1973 b).

_____, R. H. Donnay, K. Loetzsch, und H. M. Kellner: The behaviour of endosulfan in an artificial microecosystem. Farbwerke Hoechst AG. Unpublished report (1976).

Kock, A. F.: Red spider—A dangerous customer in cotton. Farming in S.A., Sept. 1965, p. 42 (1965).

Koeman, J. H., J. H. Pennings, R. Rosanto, O. Soemarwoto, P. S. Tjioe, S. Blancke, S. Kusumadinata, and P. R. Djajadiredja: Metals and chlorinated hydrocarbon pesticides in samples of fish, Sawah-duck eggs, crustaceans and molluscs. Collected in Indonesia in April and May 1972. Dept. Toxicol., Agr. Univ. Wageningen, The Netherlands (1974).

Korte, F., und M. Stiasni: Insektizide im Stoffwechsel, III Mikrosynthese von ^{14}C-markiertem Telodrin. Liebigs Ann. Chem. 656, 131 (1962).

Krczal, H.: Über einen Bekämpfungsversuch gegen die Haselnussgallmilbe (Phytoptus avellanae Nal.) mit Thiodan. Nachrichtenbl. Dtsch. Pflanzenschutzdienstes (Braunschweig) 15, 2 (1963).

_____ Untersuchungen zur Biologie und Bekämpfung der Brombeergallmilbe (*Eriophyes essigi* Hassan). Der Erwerbsobstbau 11, 239 (1969).

Lal, O. P.: Efficacy of Thiodan at different concentrations against the pupae and flies of *Phytomyza atricornis* Meigen (Diptera, Agromyzidae). Z. angew. Entomol. 69, 258 (1971).

Lavabre, E. M.: Premiéres observations sur les traitements anti-mirides appliqués sous un volume extremement réduit (ULV). Café Cacao Thé 11, 135 (1971).

Le Grice, D. S., and G. S. Marr: Fruit disease control in pineapples. Farming in S.A. 46, 9 (1970).

Levi, I., P.-B. Mazur, and T. W. Nowicki: Rapid screening method for the determination of organochlorine pesticide residues in wheat by electron capture gas chromatography. J. Assoc. Off. Anal. Chemists 55, 794 (1972).

Li, C. F., R. L. Bradley, jr., and L. H. Schultz: Fate of organochlorine pesticides during processing of milk into dairy products. J. Assoc. Off. Anal. Chemists 53, 127 (1970).

Lindquist, D. A., and P. A. Dahm: Some chemical and biological experiments with Thiodan. J. Econ. Entomol. 50, 483 (1957).

Lippold, P. C., J. S. Cleere, L. M. Massey, J. B. Bourke, and A. W. Avens: Degradation of insecticides by cobalt [60]-gamma-radiation. J. Econ. Entomol. **62**, 1509 (1969).

Liska, B. J., and W. J. Stadelman: Effects of processing on pesticides in food. Residue Reviews **29**, 61 (1969).

Litchfield, J. T., jr., and F. Wilcoxon: (Math. and nomographs). J. Pharmacol. Expt. Thera. **96**, 99 (1949).

Long, W. H., E. J. Concienne, S. D. Hensley, W. J. McCormick, and L. D. Newsom: Control of the sugarcane borer with insecticides. J. Econ. Entomol. **52**, 821 (1959).

Luessem, H., und E. Schlimme: Lokalisierung des Pflanzenschutzmittels Endosulfan im Rhein und dessen Wirkung auf Fische. Gas-Wasserfach, Wasser-Abwasser **112**, 18 (1971).

Macek, K. J., C. Hutchinson, and O. B. Cope: The effects of temperature on the susceptibility of bluegills and rainbow trout to selected pesticides. Bull. Environ. Contam. Toxicol. **4**, 174 (1969).

Magadza, C. H. D.: Field observations on the environmental effect of large-scale aerial application of endosulfan in the eradication of *Glossina morsistans centralis* Westw. in the western province of Zambia in 1968. Rhod. J. Agr. Res. **16**, 211 (1978).

Maier-Bode, H.: Properties, effect, residues and analytics of the insecticide endosulfan. Residue Reviews **22**, 1 (1968).

Manglitz, G. R., J. M. Schalk, L. W. Andersen, and K. P. Pruess: Control of the army cutworm on alfalfa in Nebraska. J. Econ. Entomol. **66**, 299 (1973).

Manske, D. D., and P. E. Coneliussen: Pesticide residues in total diet samples (VII). Pest. Monit. J. **8**, 110 (1974).

_____, and R. D. Johnson: Pesticide residues in total diet samples (VIII). Pest. Monit. J. **9**, 94 (1975).

_____ _____ Residues in food and feed—Pesticide and other chemical residues in total diet samples (X). Pest. Monit. J. **10**, 134 (1977).

Marais, J. N., and R. J. van Wyk: Pesticidal residues on tobacco. Tobacco Res. Inst. Rustenburg, S. Africa. Internal report (1976).

Martens, R.: Der Abbau von Endosulfan durch Mikroorganismen des Bodens. Schriftenr. Vers. Wasser-Boden-Lufthyg., Berlin-Dahlem, Heft **37**, 167 pp. (1972).

_____ Degradation of [8,9-^{14}C]-endosulfan by soil microorganisms. Applied Environ. Microbiol. **31**, 853 (1976).

_____ Degradation of endosulfan [-8,9-^{14}C] in soil under different conditions. Bull. Environ. Contam. Toxicol. **17**, 438 (1977).

Marti, F., J. M. Carrero, S. Planes, and J. M. Del Rivero: Ensayos de lucha contra el "badoc" del avellano. Levante Agricola **39**, 29 (1965).

Marti Fabregat, F., and J. Del Rivero: Nuevos ensayos de lucha contra el "badoc" del avellano. Boletin de Patologia Vegetal y Entomologia Agricola, p. 257 (1966).

Martin, H.: Pesticide manual: Brit. Crop Prot. Council, First ed. (1968).

McCaskey, T. A., and B. J. Liska: Effect of milk processing methods on endosulfan, endosulfan sulfate and chlordane residues in milk. J. Dairy Sci. **50**, 1991 (1967).

McDonald, S.: Laboratory evaluation on several new insecticides for control of the redbacked cutworm. J. Econ. Entomol. 65, 533 (1972).

McEven, F., A. C. Davis, H. B. Rinck, jr., and F. D. Judge: Control of the cabbage aphid in Western New York. J. Econ. Entomol. 63, 601 (1970).

McLeod, H. A., and P. J. Wales: A low temperature cleanup procedure for pesticides and their metabolites in biological samples. J. Agr. Food Chem. 20, 624 (1972).

Mehesare, V. L.: Control of rice pests. Paper presented at the plant protection seminar organized by Department of Agriculture, Govt. of Bihar, Patna, Sept. 4 (1970).

Michel, H. G.: Zum Auftreten und der Bekämpfung der Johannisbeergallmücke (*Dasyneura tetensi* Rübs.). Gesunde Pflanz. 20, 50 (1968).

Miles, J. R. W., and C. R. Harris: Insecticide residues in a stream and a controlled drainage system in agricultural areas of Southwestern Ontario. Pest. Monit. J. 5, 289 (1971).

_____ _____, and P. Moy: Insecticide residues in soil of Holland Marsh, Ontario, Canada 1972-1975. J. Econ. Entomol. 71, 97 (1978).

_____, and P. Moy: Degradation of endosulfan and its metabolites by a mixed culture of soil organisms. Bull. Environ. Contam. Toxicol. 23, 13 (1979).

Miller, G. A., and C. E. Wells: Alkaline pre-column for use in gaschromatographic pesticide residue analysis. J. Assoc. Off. Agr. Chemists 52, 548 (1969).

Mills, P. S., J. H. Onley, and R. A. Gaither: Rapid method for chlorinated pesticide residues in nonfatty foods. J. Assoc. Off. Agr. Chemists 46, 186 (1963).

Mitchell, L. C.: Separation and identification of chlorinated organic pesticides by paper chromatography. XI. A study of 114 pesticide chemicals: Technical grades produced in 1957 and reference standards. J. Assoc. Off. Agr. Chemists 41, 781 (1958).

Moffitt, H. R., E. W. Anthon, and L. O. Smith: Toxicity of several commonly used orchard pesticides to adult *Hippodamia convergens.* Environ. Entomol. 1, 20 (1972).

Moore, D. H.: Field evaluation of Thiodan as an insecticide for potatoes. J. Econ. Entomol. 52, 564 (1959).

_____ Thiodan, a promising chemical for control of the lesser peach tree borer. J. Econ. Entomol. 53, 321 (1960).

Mukerjea, T. D.: Thiodan—A broad-spectrum new insecticide. "Two and a Bud", Quarterly Newsletter from Tea Res. Assoc. Tocklai Expt. Sta. Jorhat -8, Assam/India 16, No. 2, p. 6 (1969).

Mueller, H. W. K.: Zur Bekämpfung der Erdbeermilbe (Cyclamenmilbe) *Steneotarsonemus pallidus* (Banks 1901). 5. Beitrag. Nachrichtenbl. Dtsch. Pflanzenschutzdienstes (Braunschweig) 20, 73 (1968).

_____ Erfahrungen mit neuen Wirkstoffen, insbesondere Aldicarb (Temik 10 G) bei der Bekämpfung von *Tarsonemus pallidus* Banks an Erdbeere und Gloxinie (*Sinningia hybrida*). Nachrichtenbl. Dtsch. Pflanzenschutzdienstes (Braunschweig) 23, 97 (1971).

Mullet, R. P.: Tobacco pest control. Treatments for insects and diseases in beds. Agr. Ext. Serv. Univ. Tenn., Feb. (1970).

158   H. Goebel, S. Gorbach, W. Knauf, R. H. Rimpau, and H. Hüttenbach

Mullins, D. E., R. E. Johnson, and R. I. Starr: Persistence of organochlorine insecticide residues in agricultural soils of Colorado. Pest. Monit. J. 5, 260 (1971).

Musial, C. F., M. E. Peach, and D. A. Stiles: A simple procedure for the confirmation of residues of α- and β-endosulfan, dieldrin, endrin and heptachlorepoxide. Bull. Environ. Contam. Toxicol. 16, 98 (1976).

Mustea, D., A. Barbulescu, M. Maties, V. Apetri, A. Peteanu, and V. Sandru: Chemical control of maize borer (Ostrinia nubilalis Hbn.). An. Inst. Cercet. Prot. Plant. Inst. Cent. Cercet. Agr., Bucharest 6, 253 (1970).

National Cancer Institute, National Institutes of Health, Public Health Services. U.S. Department of Health, Education, and Welfare: Bioassay of endosulfan for possible carcinogenicity. CAS No. 115-29-7, NCI CG-TR-62. Cancerogenesis Technical Report Series No. 62, 93 pp. (1978).

National Research Council of Canada: Endosulfan: Its effects on environmental quality. NRC Associate Committee of Scientific Criteria for Environmental Quality, Report No. 11. Subcommittee of Pesticides on related Compounds. Subcommittee Report No. 3. Publication No. NRCC 14098 of the Environmental Secretariat (1975).

Netto, N. D., F. de Assis Mehezes Mariconi, and F. T. M. van der Mee: Ensaio de combate à Broca do Café—Hypothenemus hampei en condicoes de campo. "Resumos" of First Brazilian Congress about insects and diseases on coffee, Victoria, Brazil, July 4-6, 9-10 (1973).

Niagara Chemical Division, FMC Corporation, Middleport, New York, U.S.A.: Acute oral LD$_{50}$ studies with endosulfan in bobwhite and Japanese quail and mallard ducks. NCT 472.62 (1972).

Niemoeller, A.: Versuch zur Bekämpfung des Frostspanners in Kirschanbaugebieten des Mittelrheins. Gesunde Pflanz. 14, 229 (1962).

Obarski, J.: Attempts of chemical control of broad bean beetle (Bruchus rufimanus Boh.) in seed plantations of broad bean. Buletyn Warzywniczy 4, 253 (1969).

_____ Der Einfluß von Thiodanbehandlung gegen Orthops sp. und Lygus sp. auf die Produktion und Keimung von Möhren- und Petersiliensamen. Rocz. Nauk. Roln. (Warzawa) Ser. E 2, 2, 69 (1972).

Odera, J. A.: Insecticidal control of Pineus sp. (Homoptera: Adelgidae) in East Africa. PANS 17, 464 (1971).

Oeser, H., S. Gorbach, and W. Knauf: Endosulfan and the environment. Giornate Fitopathologiche (Udine) 17 pp. (1971).

Olinger, L. D., and S. H. Kerr: Effects of dimethyl sulfoxide on the biological activity of selected miticides and insecticides. J. Econ. Entomol. 62, 403 (1969).

Padilla, R., and A. Ortega: Algunas observaciones sobre la biologia y el combate de la palomilla de la papa Gnorimoschema operculella, en el Bajio. Agricultura técnica México 2, 126 (1964).

Paetzold, M., and P. Burmeister: Blattgallmücken bei Obstgehölzen. Gesunde Pflanze 21, 152 (1969).

Park, P. O., J. A. Geldhill, N. Alsop, and C. W. Lee: A large-scale scheme for the eradication of Glossina morsistans Westw. in the western province of Zambia by aerial ultra-low-volume application of endosulfan. Bull. Entomol. Res. **61**, 373 (1972).

Patel, H. K., and V. C. Patel: Insecticidal control of tobacco leaf-eating caterpillar *Spodoptera (Prodenia) litura* F. Pesticides, pp. 41-43, July (1969).

_____ _____ Field testing of modern insecticides for the control of tobacco capsule-borer (*Heliothis armigera* Hb.). Indian J. Agr. Sci. **39**, 955 (1969).

_____, A. G. Patel, and M. S. Chari: Insecticidal control of mustard saw fly *Athalia lugens proxima* Klug. Pesticides **5**, 9 (1971).

Pathak, M. D.: Application of insecticides to paddy water for more effective rice pest control. Internat. Pest Control **10**, 12 (1968).

Paulino, A. E., I. P. R. Andrade, J. B. Metiello, and R. G. Abvei: Efficiencia de insecticidas no controle do "Bicho Mineiro do Café", "Resumos" of First Brazilian Congress about insects and diseases on coffee, Victoria (Brazil) July 4-6, pp. 106-107 (1973).

Perscheid, M., and K. Ballschmiter: Cyclic ketals from 2-hydroxy-methylene-hexachloro-norborn-S-enes by alkaline dechlorination. Z. Naturforsch. **28 b**, 549 (1973).

_____, H. Schlueter, und K. Ballschmiter: Aerober Abbau von Endosulfan durch Bodenmikroorganismen. Z. Naturforsch. **28 c**, 761 (1973).

Pesante, A.: L'Acarosi delle gemme del nocciolo. Bolletin del Laboratorio Sperimentale e Osservatorio di Fitopatologia, Luglio-Dicembre 1961, p. 27 (1962).

Pesticide Chemical News Guide, Washington, D.C., U.S.A., Dec. 1 (1978).

Petrova, T. M., L. A. Koshelev, and G. P. Ivanova: The use of Thiodane for protection of black currants from soil mites. Khim. Sel. Hoz. **16**, 26 (1978).

Pezzani, G. P., and C. Ruffini: Lotta pre—e post—fiorale contre *Lithocolettis* e *Cemiostoma*. L'Informatore Agrario (Verona) No. 7, p. 5 (1971).

Pezzi, A. Orientamenti di lotta contro i microlepidotteri minatori delle foglie dei fruttiferi (*Lithocolletis blancardella* F. e *Leucoptera sciatella* Zell) e osservazioni sugli sfarfallamenti. Bolletino dell'Osservazioni per la Malatti delle Plante di Bologna 2, 10 (1967-1972).

Pierard, G.: Efficacite du Thiodan contre *Stephanoderes hampei* et *Antestiopsis lineaticollis*. Bull. INEAC **11**, 59 (1962).

Pierza, H., and J. Fischer: Red scale and red spider populations in citrus as affected by insecticidal treatments. S. A. Citrus J., p. 11 Aug. (1965).

Porter, M. L., and J. A. Burke: An isolation and cleanup procedure for low levels of organochlorine pesticide residues in fats and oils. J. Assoc. Off. Anal. Chemists **56**, 733 (1973).

Powell, D. M., B. J. Laudis, C. E. Deonier, and R. Winterfield: Controlling green peach aphid on potatoes. Agrichemical West, Aug., pp. 6-8 (1969).

Pradhan, S., M. G. Jotwani, and P. Sarup: Bio-assay of insecticides—Comparative toxicity of some important insecticides to the adults of citrus psylla, *Diaphorina citri* Kuwana. (*Psyllidae: Homoptera*), a pest of citrus. Ind. J. Hort. **16**, 252 (1959).

_____ _____, and B. D. Rai: Bioassay of the relative contact toxicity of insecticides to the larvae of Bihar hairy caterpillar, *Diacrisia obliqua* Walker (*Agrotiidae:Lepidoptera*). Ind. Oilseeds J. **4**, 142 (1960).

Prakash, S., D. S. Singh, and R. Lal: Testing of pesticides against seedflower aphid, *Dactynotus carthami* (*Aphididae:Homoptera*). Ind. J. Entomol. **26** (Part 3), 300 (1964).

Pryzgodda, W.: Einige Bemerkungen zu dem Thema Pflanzenschutzmittel und Vögel. Internat. Vogelschutz, Deutsche Sektion **1**, 24 (1961).

Purohit, M. L., A. K. Khatri, S. K. Verma, and Thakur: Note on the chemical control of army-worm of rice, *Pseudaletia separata* (Wlk.) (*Noctuidae: Lepidoptera*). Ind. J. Agr. Sci. **41**, 60 (1971).

Quintana, F. J. Ensayos de control quimico y microbiologico de la isoca de la Alfalfa *Colias lesbia* (F) (*Lep. Pieridae*) en cultivos de trebol rojo y alfalfa en Balcarce. Instituto Nacional de Tecnologia Agropecuaria (INTA). Estación Experimental Agropecuaria Balcarce. Boletin Tecnico No. 68, Julio (1968).

Rajukkannu, K., K. Saivaraj, P. Vasudevan, and M. Balasubramanian: Insecticides residues in sweet potato tubers (*Ipomoea batates*). Pesticides **12**, 21 (1978).

Rammer, I. A., E. A. Kuriz, and P. E. Primer: Control of four species of aphids on deciduous fruit and nut trees with carbofuran. J. Econ. Entomol. **62**, 498 (1969).

Randolph, N. M., G. L. Teetes, and B. E. Jeter: Insecticide sprays and granules for control of the sugarcane borer on grain sorghum. J. Econ. Entomol. **60**, 762 (1967).

Rao, D. M. R., and A. S. Murty: Persistence of endosulfan in soils. J. Agr. Food Chem. **28**, 1099 (1980).

Rathore, V. S., R. K. Raghuwanshi, N. K. Sood, and N. K. Kaushik: Control of sorghum pests. PANS **16**, 358 (1970).

Rawat, R. R., Z. Singh, and K. N. Kopoov: Arthropod pests of soybean in Madhya Pradesh. Agriculture and Agro-Industries J. Sept., p. 3 (1969).

_____, S. S. Jakhmola, and H. P. Sahu: Assessment of losses of hybrid sorghum "CSH-1" to earhead caterpillars and comparison of insecticidal control. PANS **10**, 367 (1970).

Reynolds, H. T., T. R. Fukuto, and G. D. Peterson, jr.: Effect of topical application of granulated systemic insecticides and of conventional applications of other insecticides on control of insects and spider mites on sugar beet plants. J. Econ. Entomol. **53**, 725 (1960).

Ribas, C., P. Pigati, and R. R. de Almeida: Residuos de dieldrine e endosulfan em graos de café. Biologico **40**, 120 (1974).

_____, M. S. Ferreira, and N. Dias Netto: Efeito da torração sobre residuos de lindane e endosulfan en graos de café. Biologico **43**, 208 (1977).

Riemenschneider, R.: DBP 1 117 568 Prior 1960 (1960).

Rimes, G. D.: The bark beetle in West Australian pine forests. J. Agr. West Aust. **8** (3rd series), 2 (1959).

Roberts, D.: Differential uptake of endosulfan by the tissues of *Mytilus edulis*. Bull. Environ. Contam. Toxicol. **13**, 170 (1975).

_____ The assimilation and chronic effects of sub-lethal concentrations of endosulfan on condition and spawning in the common mussel, *Mytilus edulis*. Mar. Biol. **16**, 119 (1972).

Roberts, J. E., and C. B. Dominick: Virginia tobacco insect control recommendations 1972. Ext. Div., Virginia Polytech. Inst. and State Univ. Publ. 345, Rev. Jan. (1972).

Robertson, I. A. D.: Trials of insecticides to control pests attacking cotton in Eastern Tanzania 1963 to 1967. Cotton Grow. Rev. 47, 112 (1970).

———— Insecticide control of insect pests of soya bean (Glycina max. (Linnaeus)) in Eastern Tanzania. East Afr. Agr. J. 35, 181 (1969).

Roemer, D.: Thiodan, ein Fraß- und Kontaktinsektizid. Anz. Schädlingkd. 30, 174 (1957).

Rolston, L. H., J. Bagent, R. T. Brown, and W. W. Etzel: Tests compare new and old insecticides. La Agr. 13, 14 (1970).

Rueckert, W., und K. Ballschmiter: Metabolismus der Cyclodien-Insektizide Alodan und Endosulfan (= Thiodan) in Fliegen. Z. anal. Chem. 259, 188 (1972).

Rudinsky, J. A., and L. C. Terriere: Laboratory studies on the relative contact and residual toxicity of ten test insecticides to Dendroctonus pseudotsugae Hopk. J. Econ. Entomol. 52, 485 (1959).

———— ————, and D. C. Allen: Effectiveness of various formulations of five insecticides on insects infesting douglas-fir logs. J. Econ. Entomol. 58, 949 (1960).

Sarospataki, G., and G. Farkas: Applicability of acaricides against spider mites on grape vines. Növényéd. Kut. Intéz. Közl. (Budapest) 3, 117 (1969).

Satpathy, J. M., and B. Mishra: Field tests with insecticides to control jassids and fruit borer of okra (Bhendi). Pesticides 3, 27 (1969).

Schifferli, A.: Auswirkung einer Endosulfan-Behandlung gegen Maikäfer auf den Vogelbestand der betroffenen Wälder. Der Ornithologische Beobachter 64, 10 (1967).

Schlunegger, U. P.: Zur quantitativen Isolierung lipophiler Substanzen, insbesondere des Thiodans (Endosulfan) aus Organen. Arch. Toxikol. 23, 122 (1968).

Schmidlin-Mészáros, J., und E. Romann: Eine accidentelle Vergiftung von Kühen mit Endosulfan (Thiodan). Mitt. Gebiete Lebensmitteluntersuch. Hyg. 62, 110 (1971).

Schmutterer, H.: Pests of crops in Northeast and Central Africa. Gustav Fischer Verlag, Stuttgart-Portland U.S.A. (1969).

———— Schädlingsbekämpfungsprobleme im südostafrikanischen Baumwollanbau. Der Tropenlandwirt (Witzenhausen) 73, 126 (1972).

Schoettger, R. A.: 35. Investigations in fish control. Toxicology of Thiodan in several fish and aquatic invertebrates. U.S. Dept. Int., Washington, D.C. Jan. (1970).

Schulze, E.-F.: Untersuchungen zur Wirkungsweise des Insektizids Thiodan. Dissertation (Universität Würzburg) (1965).

———— Verlauf von Körperinnentemperatur und Transpiration bei Blaberus discoidalis Sv. nach der Begiftung mit Thiodan. Z. angew. Zool. 54, 91 (1967).

Schumacher, G., W. Klein, und F. Korte: Beiträge zur ökologischen Chemie. XXXII. Photochemische Reaktionen des Endosulfans in Lösung. Tetrahedron Letters 24, 2229 (1971).

Schuphan, I., und K. Ballschmiter: Zur Persistenz von Hexachlorbicyclo-[2.2.1]-hepten-Derivaten. Z. anal. Chem. **259**, 25 (1972).

———, K. Ballschmiter, und G. Toelg: Zum Metabolismus des Endosulfans in Ratten und Mäusch. Z. Naturschforsch. **23**, 701 (1968).

———, B. Sajko, and K. Ballschmiter: Zum chemischen und photochemischen Abbau der Cyclodien-Insektizide Aldrin, Dieldrin, Endosulfan und weiterer Hexachlorbicyclo-[2.2.1]-hepten-Derivate. Z. Naturforsch. **27 b**, 147 (1972).

Sharma, S. K., S. D. Mathur, R. M. Khan, and B. N. Mathur: Evaluation of some modern insecticides for the control of insect pests of cotton by means of aerial spraying and their effect on parasites and predators. Z. Pflanzenkr. Pflanzenschutz **28**, 286 (1971).

Shorey, H. H.: Effect of various insecticide treatments on populations of the green peach aphid on peppers in Southern California. J. Econ. Entomol. **54**, 279 (1961).

Sidhu, A. S., and G. Singh: Studies on the chemical control of *Oligonychus mangiferus* (Rahman and Sapra) on grape-vine. J. Res. **8**, 463 (1971).

Singh, J. P.: Effect of Thiodan against honey bees and mustard aphid, *Lipaphis erysimi* (Kalt) *(Homoptera:Aphididae)*. Proc. Beekeeping Seminar held at Almora (Utar Pradesh) on 12th and 13th Mar. 1968, under auspices of All India Bee Keepers Assoc., Poona. Res. Publ. No. 118, Coll. Agr., P. Agr. Univ. Panthagar, Distr., Nainital U.P. (1968 a).

——— Insect pests of soybean in Tarai the foot hills of the Himalayas and Kumaon hills. Proc. Conf. on Soybean Production and Marketing, Jawaharlal Nehru Krishi Vishwa. Vidyalaya Jabalpur (Madhya Pradesh), Sept. 20-22 (1968 b).

——— Die Rolle des Endosulfans in der integrierten Schädlingsbekämpfung. Gesunde Pflanz. **21**, 183 (1969 a).

——— Beware of pests of oilseed crops. Ind. Farmers' Digest **2**, 51 (1969 b).

———, and R. P. Chawla: Persistence of endosulfan on grapes. Pesticides **13**, 46 (1979).

———, and R. C. Chibber: Beet armyworm on soybean. FAO Plant Prot. Bull. **19**, 116 (1971).

———, and R. C. Shri: Insect enemies of soybean and their control. Ind. Farmers' Digest **2**, 13 (1969).

Sissons, D. J., and G. M. Telling: Rapid procedures for the routine determination of organophosphorous insecticide residues in vegetables I. Determination of hexane-soluble insecticides by gas-liquid chromatography and total-phosphorous procedures. J. Chromatog. **47**, 328 (1970).

——— ———, and C. D. Usher: A rapid and sensitive procedure for the routine determination of organo-chlorine pesticide residues in vegetables. J. Chromatog. **33**, 435 (1968).

Skrocki, C.: Bekämpfung des Kohlschotenrüßlers (*Ceutorhynchus assimilis* Payk) an Winterraps. Rocz. Nauk. Roln. (Warszawa) Ser. E **2**, 21 (1972).

Smith, D. C.: Pesticide residues in the total diet in Canada. Pest. Sci. **2**, 92 (1971).

———, R. Leduc, and C. Charbonneau: Pesticide residues in the total diet in Canada III—1971. Pest. Sci. **4**, 211 (1973).

_____ _____, and L. Tremblay: Pesticide residues in the total diet in Canada IV−1972-73. Pest. Sci. 6, 75 (1975).

_____, E. Sandi, and K. Leduc: Pesticide residues in the total diet in Canada II. Pest. Sci. 3, 207 (1972).

Sounders, J. L., and D. A. Barstow: Adelges cooley (Homoptera:Phylloxeridae) control on douglas-fir christmas trees. J. Econ. Entomol. 63, 150 (1970).

_____ _____ Trisetacus campnodus control on Pinus sylvestris. J. Econ. Entomol. 65, 500 (1972 a).

_____ _____ ULV sprays for Schizolachnus pineti control in christmas trees plantations. J. Econ. Entomol. 65, 896 (1972 b).

_____, J. K. Knoke, and D. M. Norris: Endosulfan and lindane residues on the trunk bark of Theobroma cacao for the control of Xyleborus ferrugineus. J. Econ. Entomol. 60, 79 (1967).

Spielberger, V.: Tsetse (Diptera-Glossinidae) eradication by aerial (helicopter) spraying of persistent insecticides in Nigeria. Bull. Entomol. Res. 67, 589 (1977).

Srivastava, A. S., K. M. Srivastava, and S. S. Katiyar: Bionomics and control of the rice case worm, Nymphula depunctalis, in India. Internat. Pest Control 12, 18 (1970).

_____, Y. P. Singh, R. C. Pandey, and B. K. Awasthi: Bionomics and control of the mango mealy bug. World Crops Mar./Apr., p. 87 (1973).

Sternlicht, M.: Further trials in the control of the citrus bud mite Aceria sheldoni (Ewing) (Eriphyidae, Acarina). Z. angew. Entomol. 64, 137 (1969).

Stevens, L. J., C. W. Collier, and D. W. Woodham: Monitoring pesticides in soils from areas of regular limited and no pesticide use. Pest. Monit. J. 4, 145 (1970).

Stoeva, R.: Some chemical substances for the control of the cotton bollworm (in Bulgarian). Rat. Zasht. 16, 16 (1968).

Suber, E. F., R. B. Chalfant, and T. D. Canerday: Toxicity of insecticides to the cowpea curculio in the laboratory. J. Econ. Entomol. 64, 1080 (1971).

Summers, M., D. Donaldson, and S. Togashi: Control of peach twig borer on almonds and peaches in California. J. Econ. Entomol. 52, 637 (1959).

Swirski, E., S. Amitai, S. Greenberg, and N. Dorzia: Field trials on the toxicity of some carbamates and endosulfan to predaceous mites (Acarina, Phytoseiidae). Israel J. Agr. Res. 18, 41 (1968).

Tappan, W. B., C. H. van Middelem, and H. A. Moye: DDT, endosulfan and parathion residues on cigar-wrapper tobacco. J. Econ. Entomol. 60, 765 (1967).

Tessari, J. D., and D. L. Spencer: Air sampling for pesticides in the human environment. J. Assoc. Off. Anal. Chemists 54, 1376 (1971).

Thian Hua, H.: The bean-fly (Melanagromyza phaseoli Coq.) and experiments on its control. Malaysian Agr. J. 46, No. 2 (1967).

Thier, W.: Teil I: Analysengang zur Ermittlung von Pesticidrückständen in Pflanzenmaterial. Deutsche Lebensmittelrundschau 68, 345 (1972); Teil II. 68, 397 (1972).

Thorbecke, H. J., and J. Fisher: New chemical development for insect control. S. A. Citrus J., p. 25 (1963).

Tripathi, R. L.: Relative toxicity of some insecticides to the hairy caterpillar. Ind. J. Entomol. **28** (Part IV), 561 (1966).

Uk, S., and Ch. M. Himel: Gas chromatographic method for analysis of chlorpyrifos and endosulfan insecticides in topically treated houseflies. J. Agr. Food Chem. **20**, 638 (1972).

U.S. Department of Health, Education and Welfare, Food and Drug Administration: Pesticide analytical manual, Section 211 (1971, 1972).

Valmayor, R. W.: Important diseases and pests and their control. Agroservice, Esso, Bull. No. 16, May-June, 19 pp. (1968).

Van Den Bruel, W. E.: Le problème du tarsoneme du fraisier *Tarsonemus pallidus* Banks. Commentaires sur les possibilités de destruction de l'acarien dans les champs (situation 1958). Rev. l'Agriculture **13**, 9 (1960).

Van De Vrie, M.: The biology and control of the black currant gall mite *Cecidophyopsis ribis*. Neth. Pl. Path. **73**, 170 (1967).

Van Dyk, L. P., and C. G. Greeff: Endosulfan pollution of rivers and streams in the Loskop Dam Cotton-Growing Area. Agrochemophysica **9**, 71 (1977).

_____, B. C. Breedt, and P. R. de Beer: Die Bepaling van Organochloor-Plaagdoderresidu's in Suid-Afrikaanse Hoendereiers. Agrochemophysica **10**, 47 (1978).

Varlet, G.: Une "Operation-Hannetons" en 1972: L'évolution des methodes de lutte. Phytoma **25**, 9 (1973).

Verband der Zigarettenindustrie, Wissenschaftliche Abteilung, Hamburg BRD: Private Mitteilung (1975).

Voigt, J., und R. Noske: Beitrag zur Thiodan-analytik. Die Nahrung **12**, 391 (1968).

Waede, M.: Versuche zur Bekämpfung der Kohlschotenmücke (*Dasyneura brassicae* Winn.) in blühenden Oelfruchtbeständen mit Hilfe des Kaltnebelverfahrens. Nachrichtenbl. Dtsch. Pflanzenschutzdienstes (Braunschweig) **12**, 65 (1960).

_____ Die Bewährung des Kaltnebelverfahrens bei einem Großeinsatz zur Bekämpfung der Kohlschotenmücke (*Dasyneura brassicae* Winn.). Nachrichtenbl. Dtsch. Pflanzenschutzdienstes (Braunschweig) **13**, 70 (1961).

Waffelaert, P.: Nouvelle perspective de lutte contre le puceron lanigere. Bull. Technique No. 21, 20 (1962).

Walsh, A.: The pathology of pesticide poisoning in fish. Diss. Abstr. Internat. **B35**, 3423 (1975).

Watson-Cook, D.: Notes on aerial spraying against tsetse flies. Unpublished report (1973).

Wegmann, R. C. C., and P. A. Greve: Organochlorines, cholinesterase inhibitors and aromatic amines in dutch water samples, September 1969-December 1975. Pest. Monit. J. **12**, 149 (1978).

Weidenmüeller, H.-L., M. Mracek, und H. Oeser: Abbau von Thiodan durch Mikroorganismen. Internal Report, Farbwerke Hoechst AG (1971).

Weinmann, W.-D.: Analysenmethode zur Bestimmung von α- und β-Endosulfan in technischen Wirkstoff und seinen Formulierungen. Nachrichtenbl. Dtsch. Pflanzenschutzdienstes (Braunschweig) **22**, 24 (1970).

Weiss, E. A.: Dwarf castor, a promising crop for East Africa. World Crops **18**, No. 4, 43 (1966).

Werkhof GmbH, Hamburg BRD: Persönliche Mitteilung (1976).

Wheatley, P. E.: The giant coffee looper, *Ascotis selenaria reciprocaria* Walk. (*Lepidoptera:Geometridae*). East Afr. Agr. For. J., Oct., p. 143 (1963).

Wheeler, W. B., D. E. H. Frear, R. O. Mumma, R. H. Hamilton, and R. C. Cotner: Quantitative extraction of root-absorbed dieldrin from the aerial parts of crops. J. Agr. Food Chem. 15, 227 (1967).

WHO: 1967 Evaluations of some pesticide residues in food. Food and Agriculture Organization of the United Nations, World Health Organization, Rome, p. 134 (1968).

———— 1968 Evaluations of some pesticide residues in food. Food and Agriculture Organization of the United Nations, World Health Organization, Geneva, p. 148 (1969).

———— 1971 Evaluations of some pesticide residues in food. WHO Pesticide Residues Series No. 1, World Health Organization, Geneva, p. 98 (1972).

Wood, B. J.: Developments in oil palm pest management: 1965-71. Planter (Kuala Lumpur) 48, 93 (1972).

Yadav, T. D., and A. Varma: Evaluation of pesticides against mango bud mite, *Aceria mangiferae* Hassan (*Eriophyidae:Acarina*). Ind. J. Entomol. 31, 244 (1969).

Yap, H. H., D. Desaiah, L. H. Cutkomp, and R. B. Koch: In vitro inhibition of fish brain ATPase activity by cyclodiene insecticides and related compounds. Bull. Environ. Contam. Toxicol. 14, 163 (1975).

Young, J. R., and L. P. Ditman: The effectiveness of some insecticides on several vegetable crops. J. Econ. Entomol. 52, 477 (1959).

Young, S. J. V., and J. A. Burke: Micro scale alkali treatment for use in pesticide residue confirmation and sample cleanup. Bull. Environ. Contam. Toxicol. 7, 160 (1972).

Yun-Pei Sun: Correlation of toxicity of insecticides to the house fly and to the mouse. J. Econ. Entomol. 65, 632 (1972).

Zaets, V. G.: Mites on currant (in Russian). Zashch. Rast. 13, 46 (1968).

Zeid, M., A. A. Saad, A. M. Ayad, G. Tansawi, and M. E. Eldefvawi: Laboratory and field evaluation of insecticides against Egyptian cotton leafworm. J. Econ. Entomol. 61, 1183 (1968).

Zioni, E.: Le psille e la loro lotta. Note techniche di agricultura del Instituto Federale di Credito Agrario per il Piemonte e la Ligura, Allesandria (1972).

Zyl, H. V.: Black spot in pineapples. Breakthrough registered in control of the disease. Farmers Weekly, 24 Sept. (1969).

Manuscript received March 30, 1981; accepted August 18, 1981.

Subject Index

Acanthomia sp. 123
Acaphylla sp. 133
Aceria sp. 131, 132
Achaea spp. 122, 126
Acrosternum sp. 127
Acrythosiphon sp. 122
Acyrtosiphon sp. 122
Adelges sp. 136
Aedes sp. 43
Aelia sp. 120
Agriotes sp. 120
Agromyza sp. 122
Agrotis sp. 121, 128
Alabama sp. 123
Aldrin 22
Aleurocanthus sp. 132
Alfalfa 75, 118
Almonds 114, 118
Amblyseius sp. 47
Amsacta sp. 126
Anarsia spp. 122, 130
Anas sp. 51
Ancylus sp. 43
Anomis sp. 125
Anser sp. 51
Antestiopsis sp. 132, 133
Anthonomus spp. 123, 127, 130
Anticarsia sp. 122
Antigastra sp. 126
Anuraphis sp. 130
Aonidiella sp. 47, 132
Aphelinus sp. 47
Aphidula sp. 129
Aphis spp. 122–124, 126, 129, 130
Aphytis spp. 47, 132
Apion spp. 125, 135
Apples 75, 114, 118, 129
Apricots 86, 87, 114, 118
Apsylla sp. 131
Artemia sp. 43
Artichokes 114, 118
Ascotis spp. 131–133
Asellus sp. 43
Asparagus 114
Aspergillus spp. 30
Athalia sp. 126
Aulacorthum sp. 96

Avocados 114
Azinphosmethyl 48, 94, 100

Bacillus sp. 29
Bacterium spp. 20
Balaninus sp. 131
Bananas 114
Banol 100
Baris sp. 134
Barley 114, 118
Bathycoelia sp. 133
Batrachedra sp. 131
Bay 78182 100
Bay 93820 100
Beans 52, 114, 118, 122
Beef meat and products 118
Beets 117
Bemisia sp. 122, 124, 127, 128, 132
Benlate 57
Beosit 8
Beta sp. 52
BHC 13
Binapacryl 48
Biston sp. 133
Bixadus sp. 133
Blaberus spp. 98, 104, 106, 108–110
Blackberries 114, 128
Black salsify 114
Blood 15
Blueberries 114, 118
Bobwhite 51
Botrytis sp. 28
Brassica sp. 99
Brevicoryne sp. 127
Broad beans 114
Broccoli 114, 118
Bromophos 50
Bruchus sp. 122
Brumus spp. 47, 49
Brussels sprouts 114, 118
Buprestis sp. 135
Busseola spp. 113, 120, 121
Butter 17

Cabbages 75, 114, 118, 127
Calacorus sp. 133
Calliphora sp. 105

Calotermes spp. 104, 105, 107, 111, 136
Cambarus sp. 43
Capsicum sp. 96
Captan 57
Carassius spp. 45, 71, 72
Carbaryl 48, 50, 100
———— persistence in water 68
Carbofuran 94
Carbophenothion 13
Carrots 114, 118, 120, 128
Cashew nuts 114
Castor-oil plants 126
Catastomus spp. 45, 71
Cauliflower 114, 118
Cecidophyopsis sp. 128
Celery 114, 118
Cemiostoma sp. 130
Cephonodes sp. 133
Cerambycidae sp. 135
Ceratitis sp. 90
Ceratocystis sp. 133
Ceutorhynchus sp. 125
Chalcodermus spp. 93, 94, 122
Champignons 114
Channa sp. 46
Chanos sp. 73
Chard 114
Cheimatobia sp. 130
Cherries 114, 118
Chestnuts 114
Chicory 115
Chilo sp. 120, 121, 134
Chilotraea sp. 121, 134
Chironomidae sp. 43
Chironomus sp. 43
Chlorella spp. 28, 39, 55
Chromaphis sp. 131
Chrypsopa sp. 49
Chrysopa sp. 47
Cirphis sp. 121
Citrus 115, 131
Cladophora sp. 39
Cloeon sp. 43
Clover 75, 135
Clysia sp. 129
Cnaphalocrocis sp. 121
Cocoa 133
Coconuts 115
Coffee 76 ff., 132
Colias sp. 135
Colinus sp. 51
Collards 114, 118
Contarinia spp. 121, 136
Copepoda sp. 43
Corixidae sp. 43
Corn 75, 115, 117, 118, 120

Corynebacterium sp. 29
Cotton 43, 47, 76, 86, 123
Cottonseed and oil 115, 118, 120
Coturnix sp. 51
Crabs 73
Cryptomyzus sp. 129
Cucumbers 115, 118
Currants 76, 115, 128
Cyclodan 8
Cyclops sp. 43
Cyprinus spp. 45, 73
Cyrtopeltis sp. 128

Dactynotus sp. 126
Dacus spp. 90
Daphnia spp. 43, 72, 110
Darna sp. 134
Dasychira sp. 122
Dasyneura spp. 125, 126, 128
Dasynus sp. 135
Date palms 131
Dates 115
DDD, *see* TDE
DDE 20
DDT 13, 18, 20, 48, 57, 74, 79, 88–90, 100, 102, 124, 125
Dendroctonus spp. 135, 136
Derris 50
Diacrisia spp. 122, 125, 126
Dialeurodes sp. 132
Diaphorina spp. 131, 132
Diatraea spp. 120, 134
Diazinon 13, 48, 50
Dichocrocis sp. 133
Dicofol 88
Diconocoris sp. 135
Dieldrin 57, 74, 79
———— persistence in water 68
Dimethoate 47, 49, 50, 125
———— persistence in water 68
Diparopsis spp. 123, 124
Diplosis sp. 134
Diprion spp. 104–107, 109, 110
Distantiella sp. 133
Doralis sp. 96
Drosicha sp. 131
Ducks 51, 73
Dursban 94

Earias sp. 123–125, 127
Eggplants 115, 118
Eggs 18
Empoasca sp. 122–124, 126, 127, 133
Endosulfan 1 ff.
———— acetylation 14

———— acute oral toxicities 38
———— adsorption 54
———— adsorption isotherms 54, 55
———— analytical methods 13 ff.
———— analytics 1 ff.
———— and body temperature 107
———— and insect body inflation 110
———— and oxygen consumption 105 ff.
———— and pH of hemolymph 110
———— and pulse rate 109
———— and pyknoses of nuclei 111
———— and water turnover 108
———— application rates 113 ff.
———— average daily intake 85
———— biochemical degradation 56
———— bioconcentration 71
———— carcinogenicity test 38
———— catalyzed hydrolysis 54
———— chemical degradation 10
———— chemical designation 6
———— cleanup (see also Endosulfan, analytical methods) 17
———— colorimetry 20
———— compounds, acute oral toxicities 38, 40, 41
———— compounds, GLC columns 22, 24–27
———— compounds, GLC retention times 24–27
———— compounds, residue persistences 53
———— compounds, TLC R_f values 25, 27
———— concentration factor 74
———— contact effect 89 ff.
———— control of tsetse fly 137
———— degradation scheme 34, 36
Endosulfan diol, acute oral toxicity 41
———— diol, isolation 23
Endosulfan dusts 113
———— effect on energy-dependent processes 46
———— effect on fish 44, 45
———— effect on fish food organisms 43
———— effect on honeybees 50
———— effects 1 ff.
———— effects on arthropods 89 ff.
———— effects on beneficial organisms 47
———— effects on birds 51
———— effects on mammals 52
———— emulsifiable concentrate 113
———— environmental behavior 52 ff.
———— environmental toxicology 38 ff.
Endosulfan ether, acute oral toxicity 40
Endosulfan excretion 30 ff.
———— excretion by fish 71, 72

———— external intoxicating symptoms 104
———— extraction 15 ff.
———— extraction efficiency vs. water content of sample 16
———— extraction from air 17
———— extraction from butter 17
———— extraction from blood 15
———— extraction from eggs 18
———— extraction from hay 18
———— extraction from water 17
———— feed effect 96 ff.
———— fields of application 112 ff.
———— formulation analysis 15
———— formulations 8, 113
———— GLC 13, 15, 18
———— granules 113
———— half-life in houseflies 74
———— half-lives in water and pH 54
———— HPLC 21
———— hydrolysis 13, 54
Endosulfan hydroxyether, acute oral toxicity 41
Endosulfan in alfalfa 75
———— in apples 75, 85
———— in beans 85
———— in cabbages 75
———— in canned apricots 86, 87
———— in canned spinach 86, 87
———— in clover 75
———— in coffee and coffee infusions 76 ff.
———— in coffee, effect of roasting 77
———— in corn 75
———— in cotton and cotton fractions 76
———— in currants 76
———— in daily food 83
———— in fish tissues 71
———— in forestry 135
———— in grapes 75
———— in houseflies, half-life 30
———— in laboratory ecosystems 74
———— in lettuce 76
———— in milk 30, 34
———— in milk, tolerance 112
———— in oil palm 75
———— in onions 75
———— in pickles 86
———— in processed cotton 86
———— in processed grapes 86
———— in processed soybeans 86
———— in processed sugarbeets 86
———— in rapeseed 75
———— in rat tissues 35
———— in ready-to-serve foods 85
———— in rice paddy water 42

———— in river water 43 ff.
———— in soil, metabolic pathway 31
———— in soils 58–67
———— in spinach 75
———— in sugarbeets 75
———— in sugarcane 76
———— in sweet potatoes 76
———— in tea and tea infusions 76
———— in terrestrial environments 74
———— in the Main River 70
———— in the Rhine River 69
———— in tobacco 79 ff.
———— in tobaccos of different origins 83
———— in tobacco smoke 83
———— in tomatoes 75
———— intoxication phases 104
———— intoxication symptoms, insects 106
———— in water, destructive oxidation 70, 88
———— IR assay 13
Endosulfan isomers 5 ff.
———— isomers, effects 101 ff.
———— isomers, GLC retention times 22
———— isomers, volatilities 53
Endosulfan lactone, acute oral toxicity 41
Endosulfan metabolism 27 ff.
———— metabolism in animals 30 ff.
———— metabolism in insects 30
———— metabolism in microorganisms 28 ff.
———— metabolism in plants 27
———— metabolites, acute oral toxicities 38
———— metabolites, analysis 23 ff.
———— metabolites, analytical methods 13 ff.
———— metabolites, cleanup 24
———— metabolites, effects 101 ff.
———— metabolites, effects on arthropods 89 ff.
———— metabolites, environmental toxicology 38 ff.
———— metabolites, extraction 23
———— metabolites, GLC 24
———— metabolites, half-lives in soil 30, 33
———— metabolites, HPLC 26
———— metabolites, TLC 25
———— metabolites, toxicity 37 ff.
———— metabolites, toxicity to algae 39
———— metabolites, toxicity to animals 47
———— metabolites, toxicity to aquatic organisms 39
———— metabolites, toxicity to bacteria 39

———— metabolites, toxicity to crayfish 42
———— metabolites, toxicity to fish 44
———— metabolites, toxicity to insects 42
———— metabolites, toxicity to invertebrates 47 ff.
———— metabolites, toxicity to molluscs 42
———— metabolites, toxicity to plants 47
———— metabolites, toxicity to soil fungi 39
———— metabolites, toxicity to vertebrates 51
———— mutagenicity test 38
———— NMR assay 13
———— no-effect level 37
———— on apples 129
———— on beans 122
———— on blackberries 128
———— on cabbages 127
———— on carrots 128
———— on castor-oil plants 126
———— on citrus 131
———— on clover 135
———— on cocoa 133
———— on coffee 132
———— on corn 120
———— on cotton 123
———— on currants 128
———— on date palms 131
———— on hazelnuts 130
———— on jute 125
———— on lucerne 27, 135
———— on mangos 131
———— on mulberry trees 134
———— on mushrooms 127
———— on mustard seed 125
———— on oil palms 134
———— on okra 127
———— on olives 134
———— on onions 127
———— on parsley 128
———— on peaches 130
———— on peanuts 126
———— on pears 129
———— on peas 122
———— on pepper 135
———— on peppers 126
———— on pineapples 129
———— on potatoes 123
———— on rape 125
———— on rice 121
———— on rubber trees 134
———— on safflower 125
———— on sorghum 120
———— on soybeans 122

———— on strawberries 129
———— on sugarbeets 123
———— on sugarcane 134
———— on sunflower 126
———— on tea 133
———— on tobacco 28, 29, 128
———— on tomatoes 126
———— on vines 129
———— on walnuts 130
———— on wheat 120
———— oxidation and oxidative
degradation 30, 32, 55
———— persistence in water 68–70
———— photochemical degradation 10
———— photodegradation 56
———— physicochemical properties 8
———— physiological responses 105 ff.
———— phytotoxicity 47
———— properties 1 ff.
———— qualitative determination 13 ff.
———— quantitative determination 15 ff.
———— radiolabeled 7
———— reduction 13
———— residue analysis 15 ff.
———— residue analytical methods 13 ff.
———— residue values 52 ff.
———— residues 1 ff.
———— residues, confirmation 14
———— residues, effects of processing on
86 ff.
———— residues in food (see also specific
foods) 74 ff.
———— residues in soil 56 ff.
———— residues in stimulants 74 ff.
———— residues in water 57 ff.
———— residues, recommended analytical
methods 21 ff.
Endosulfan sulphate, acute oral toxicity
40
———— sulfate, analysis 23
———— sulfate in milk 23, 74, 88
———— sulfate in processed milk products
88
———— sulfate in soils 67
———— sulfate in tobacco 79
———— sulfate in water 70
———— sulfate on alfalfa, persistence 28
Endosulfan, synthesis 6
———— systemicity 27
———— technical grade product analysis
15
———— teratogenicity test 37
———— three-generation reproduction
study 37
———— TLC 20
———— tolerances 52, 112 ff.

———— toxicity 37 ff.
———— toxicity to algae 39
———— toxicity to animals 47
———— toxicity to aquatic organisms 39
———— toxicity to bacteria 39
———— toxicity to crayfish 42
———— toxicity to fish 44
———— toxicity to insects (see also
specific insects) 42
———— toxicity to invertebrates 47 ff.
———— toxicity to molluscs 42
———— toxicity to plants 47
———— toxicity to soil fungi 39
———— toxicity to vertebrates 51
———— transformation products 9
———— two-year feeding test 37
———— ULV 113
———— uptake by animals and plants
71 ff.
———— uptake from water 71 ff.
———— volatility 52
———— wettable powder 113
Endrin 73, 124
Ephemeroptera sp. 43
Epicampoptera sp. 133
Epilachna spp. 92, 93, 122
Epitrimerus sp. 129
Epitrix spp. 123, 127
Erannis sp. 130
Eriophyes sp. 129
Eriosoma spp. 47, 130
Esox sp. 45
Etiella sp. 122
Eugnamptus sp. 131
Euproctis sp. 126, 130, 133
Eurystylus sp. 126
Euschistus sp. 122
Euxoa spp. 99, 101, 121, 135

Fennel greens 115
Filberts 118
Flax seeds 115
Formothion 47, 49
Frankliniella sp. 128

Garlic 115
Geese 51
Gnathotrichus spp. 135, 136
Goat meat and products 118
Gobio sp. 45
Goldfish 71, 72
Golden orfe 45
Gooseberries 115
Grapes 75, 86, 115, 118
Green beans 115

Habrochila spp. 132, 133
Haltica spp. 123, 129
Hampala sp. 73
Hay 18
Hazelnuts 115, 130
Heliothis spp. 99, 121, 123, 124, 126, 128, 132
Helopeltis spp. 124, 133
Hemitarsonemus spp. 124, 125, 133, 134
Heptachlor 13, 88
Heptachlor epoxide 13
Hippodamia sp. 48
Hispa sp. 121
Holotrichia spp. 133, 134
Homeosoma sp. 126
Horse meat and products 118
Horseradish 115
Hyalopterus sp. 130
Hydrianidae sp. 43
Hydrophylidae sp. 43
Hylemya sp. 120
Hylobius sp. 136
Hylurgops sp. 136
Hypera sp. 135
Hyponomeuta sp. 130
Hypothenemus sp. 132, 133

Idiocerus spp. 131
Idus sp. 45
Iguanas sp. 46
Imidan 100
Insect body temperatures, reaction types 107, 108
Ips spp. 136
Ischnura sp. 43

Jute 125

Kale 115, 118
Kingfisher 52
"k.o. phase" 105

Lasius sp. 98
Lebistes spp. 45, 72
Leeks 115
Leiognathus sp. 73
Lepomis sp. 46
Leptinotarsa sp. 91, 92, 94, 98–100, 123, 127
Leptispa sp. 122
Leptocorisa sp. 121
Lettuce 76, 115, 118
Leucaspius sp. 45
Limnaea sp. 43
Lindane 13, 50, 88
Liothrips sp. 134

Lipaphis sp. 126
Lipids, freezing-out technique 17
Lithocolletis sp. 130
Longitarsus sp. 122
Lophobaris sp. 135
Loxostege spp. 123
Lucerne 27, 52, 56, 135
Lygus spp. 123, 124, 128, 133
Lymantra sp. 130
Lymantria spp. 104–107

Macadamia nuts 115, 118
Macrones sp. 73
Macrosiphum spp. 126, 127
Maize, *see* Corn
Malacosoma sp. 130
Malathion 13
Malix 8
Mamestra sp. 123, 127
Mancozeb 48
Manduca spp. 126, 128
Maneb 48
Mangos 115, 131
Market basket analyses 21
Melanagromyza spp. 122, 123
Melanitis sp. 122
Meligethes sp. 125
Melolontha sp. 135
Melons 115, 118
Methomyl 100
Methoxychlor 20, 50
Methyldemeton 47, 49
Methyl parathion 13, 125
Mevinphos 94
Microtermes spp. 131, 133, 135
Millet 116
Milk 23, 30, 34, 74, 88, 115, 118, 120
Molluscs 73
Monochamus sp. 136
Monuron, persistence in water 68
Mulberry trees 134
Musca spp. 30, 89, 90, 101, 102
Mushrooms 116, 127
Mussels 72
Mustard greens and seeds 116, 118, 125
Mycobacterium sp. 29
Myllocerus sp. 124
Mystacoleucus sp. 73
Mytilus spp. 44, 72
Myzus spp. 96, 122, 123, 127, 128, 130

Nectarines 116, 118
Nemourella sp. 43
Nephotettix sp. 121
Nezara spp. 122, 123, 126, 127, 135
Nilaparvata sp. 95, 96

Notonectidae sp. 43
Numicia sp. 135
Nymphula sp. 121
Nysius sp. 126

Oats and oat straw 116, 118
Oberea sp. 122
Odontotermes sp. 131, 133, 135
Oecophylla sp. 131
Oil palms 75, 134
Okra 127
Oligonychus sp. 131
Olives 116, 134
Omnatissus sp. 131
Onchorhynchus sp. 46
Oncopeltus sp. 99, 100
Onions 75, 116, 120, 127
Oregma sp. 134
Orthops sp. 128
Orthotomicus sp. 136
Oscinella spp. 120, 122
Ostracoda sp. 43
Ostrinia sp. 120
Ovex 13

Pangasius sp. 73
Panonychus sp. 119
Papilio spp. 131, 132
Paprika 116
Papularia sp. 29
Parasa sp. 133
Paratelphusa sp. 73
Parathion 13, 47, 48, 50, 94
——— persistence in water 68
Parsley 116, 128
Parus sp. 51
PCBs 13
Peaches 116, 118, 130
Peanuts 116, 126
Pears 116, 118, 129
Peas 116, 118, 122
Pecans 116, 118
Pectinophora sp. 123
Pegomya sp. 123
Pelopidas sp. 122
Penicillium spp. 28, 29
Peppers 126, 135
Peregrinus sp. 121
Perigea sp. 126
Perileucoptera sp. 133
Periplaneta spp. 94–96, 98, 104, 106, 107, 110
Perthane 13, 20
Phaseolus sp. 52
Pheasants 51
Philonthus spp. 48, 50

Phoridae sp. 127
Phormidium spp. 42
Phthorimaea sp. 123
Phyllobius sp. 130
Phyllocoptes sp. 129
Phyllodromia sp. 94, 95, 102, 103
Phyllotreta sp. 122
Physa sp. 43
Phytomyza sp. 90, 91
Phytonomus sp. 135
Phytoptus sp. 130, 131
Pickles 86
Pieris spp. 99, 127
Pineapples 116, 118, 129
Pineus sp. 136
Pissodes spp. 136
Pistachios 116
Pityogenes sp. 136
Planorbis sp. 43
Platyedra sp. 124
Plums 116, 118
Plusia spp. 122, 126–128
Plutella spp. 99, 127
Polychrosis sp. 129
Poppy seeds 116
Potatoes 116, 119, 120, 123
Pork meat and products 118
PP 511 100
Prays spp. 131, 134
Proceras sp. 134
Processing, effects on endosulfan residues 86 ff.
Prodenia sp. 99
Protoparce sp. 128
Prunes 116, 119
Pseudaletia sp. 121
Pseudomonadaceae sp. 30
Pseudoplusia sp. 122
Psylla spp. 130
Pugria sp. 122
Pumpkins 116, 119
Puntiu sp. 45
Pyrausta spp. 96, 97
Pyrethrum 50
Pyrilla sp. 134

Quail 51

Radishes 116, 117
Rape and rapeseed 48, 75, 117, 119, 125
Rasbora sp. 45
Rhopalosiphum sp. 121, 127
Rhubarb 117
Rice 68, 69, 117, 120, 121
Ronnel 13
Rubber trees 134

Rutilus sp. 45
Rye and rye straw 117, 119

Safflowers and seed 117, 119, 125
Sahlbergella sp. 133
Salmo sp. 45
Salvelinus sp. 46
Sanninoidea sp. 130
Sappaphis sp. 129
Saratherodon sp. 45
Saurida sp. 73
Scenedesmus spp. 28, 55
Schizaphis sp. 121
Schizolachnus sp. 136
Sciara sp. 127
Scirpophaga sp. 134
Scirtothrips spp. 131–134
Selenastrum sp. 39
Selenocephalus sp. 122
Sesamia sp. 121, 134
Setora sp. 134
Sheep meat and products 119
Shrimp 73
Simuliidae sp. 43
Sinapis sp. 99
Sitona sp. 122
Sitophilus sp. 92, 102, 104–108
Sogatella sp. 121
Solubea sp. 121
Sorghum 120
Soybeans 86, 122
Sparganothis sp. 129
Spices 117
Spinach 75, 86, 87, 117, 119
Spodoptera sp. 120–123, 125–128
Squash 116, 119
Steneotarsonemus sp. 129
Strawberries 117, 119, 129
Streptomyces spp. 39
Suckers 71
Sugarbeets 52, 53, 56, 67, 75, 86, 117, 119, 123
Sugarcane 76, 116, 119, 134
Sunfish 46
Sunflower seeds 117, 119, 126
Supracide 94
Sweet potatoes 76, 117, 119, 120
Syllepta sp. 123
Synanthedon spp. 130

Tachyporus spp. 48, 50
Tarsonemus sp. 129

TDE 20, 21
Tea 76, 117, 119, 120, 133
Temik 57
Temnorhinus sp. 123
Tenebrio sp. 104, 106
Tetanychus sp. 119
Tetradifon 13
Therioaphis sp. 135
Thifor 8
Thimul 8
Thiodan, *see* Endosulfan
Thiometon 47, 49
Thosea spp. 133, 134
Thrips sp. 124, 126–128
Tilapia sp. 73
Tirathaba sp. 134
Tobacco 15, 16, 18, 20, 28, 29, 79 ff., 128
Tomatoes 75, 117, 119, 126
Tomicus sp. 136
Toxaphene 50, 88, 94
Toxoptera spp. 132, 133, 135
Trialeurodes sp. 127
Triazophos 125
Trichogaster sp. 73
Trichoplusia sp. 123, 127, 128
Trioza sp. 132
Trisetacus sp. 136
Trypanosoma spp. 137
Tryporyza sp. 121
Tsetse flies 51, 137
Turdus sp. 51
Turnips and greens 117, 119
Tychius sp. 135
Typhlocyba sp. 131

Udang sp. 73
Upeneus sp. 73

Vasates sp. 127, 130
VCS 506 100
Vicia sp. 99, 122
Vines 129

Walnuts 117, 119, 130
Watercress 117, 119
Wheat and wheat straw 117, 119, 120

Xiphophorus sp. 45
Xyleborus sp. 133

Youngberries 117

INFORMATION FOR AUTHORS

RESIDUE REVIEWS

(A BOOK SERIES CONCERNED WITH RESIDUES OF PESTICIDES AND OTHER CONTAMINANTS IN THE TOTAL ENVIRONMENT)

Edited by

Francis A. Gunther

Published by

Springer-Verlag New York • Heidelberg • Berlin

The original (ribbon) copy and one good xerox or other copy of the manuscript, complete with figures and tables, are required. Manuscripts will normally be published in the order in which they are received, reviewed, and accepted. They should be sent to the editor:

Professor Francis A. Gunther
Department of Entomology
University of California
Riverside, California 92502
Telephone: (714) 787-5804/5810 (office)
(714) 688-6666 (home)

1. Manuscript

The manuscript, in English, should be typewritten, double-spaced throughout, on one side of 8½ x 11 inch blank white paper, with at least one-inch margins. The first page of the manuscript should start with the title of the manuscript, name(s) of author(s), with author affiliation(s) as first-page starred footnotes, and "Contents" section. Pages should be numbered consecutively in arabic numerals, including those bearing figures and tables only. In titles, in-text outline headings and subheadings, figure legends, and table headings only the initial word, proper names, and universally capitalized words should be capitalized.

Footnotes should be inserted in text and numbered consecutively in the text using arabic numerals.

Tables should be typed on separate sheets and numbered consecutively within the text in roman numerals; they should bear a descriptive heading, in lower case, which is underscored with one line and which starts after the word "Table" and the appropriate roman numeral; *footnotes in tables* should be designated consecutively within a table by the lower-case alphabet. *Figures* (including photographs, graphs, and line drawings) should be numbered consecutively within the text in arabic numerals; each figure should be affixed to a separate page bearing a legend (below the figure) in lower case starting with the term "Fig." and a number.

2. Summary

A concise but informative summary (double-spaced) must conclude the text of each manuscript; it should summarize the significant content and major conclusions presented. It must not be longer than two 8½ × 11 inch pages of double-spaced typing. As a summary, it should be more informative than the usual abstract.

3. References

All papers, books, and other work cited in the text must be included in a "References" section (also double-spaced) at the end of the manuscript: If comprehensive papers on the same subject have been published, they should be cited but only for exceptional reasons should the bibliographic citations extend farther back than to these papers.

The references used *in the text* should consist of the COMPLETELY CAPITALIZED author's or authors' last name(s) where one or two authors are concerned; should there be more than two authors, only the first is named and *"et al."* is added. The publication year in parentheses should follow the name. If more than one paper by one author published in the same year is cited, the letters a, b, c, etc., should follow the year, e.g., "MEIER (1958 a) found . . .", or "This method is nonspecific (MEIER 1958 a)."

In the References section, the papers cited should appear in alphabetical order according to the last name of the first author; if more than one paper by an author or authors published in the same year is cited, the papers should be listed according to the year of publication followed by a, b, c, etc., as necessary. Papers published in periodicals should be cited with COMPLETELY CAPITALIZED names and initials of all authors, together with the *full title of the paper and preferably in its original language*, title of the periodical (abbreviated in accordance with *Chemical Abstracts'* "List of Periodicals Abstracted"), number of the volume (wavy underlined), initial page, and the year in parentheses. References to unpublished papers that have been submitted for publication should be cited in the same manner as other papers except the abbreviated journal name is followed by the words "In press" or "Accepted for publication" and the year in parentheses; personal communications are to be cited similarly.

In text and in the References section, citation of *governmental agencies, educational and research institutions and foundations, professional associations, and industrial companies* should consist of the full name as used by the organization completely underscored with one line and with initial capital letters only, followed by the appropriate reference information as specified above.

Examples:

EDWARDS, C. A., and E. B. DENNIS: Some effects of aldrin and DDT on the soil fauna of arable land. Nature 188, 767(1960).

GUNTHER, F. A., J. H. BARKLEY, and W. E. WESTLAKE: Worker environment research. II. Sampling and processing techniques for determining dislodgable pesticide residues on leaf surfaces. Bull. Environ. Contam. Toxicol. Accepted for publication (1974).

HESSLER, W.: Eine einfache Nachweismethode für Paraffin in Wachsgemischen. II. Mitt. Fette, Seifen, Anstrichmittel 58, 602(1956).

MELZER, H.: The qualitative and quantitative colorimetric determination of captan. Nachrbl. deut. Pflanzenschutzdienst 14, 193(1960).

Shell Chemical Co.: Letter to EPA's "Hazardous Materials Advisory Committee," Oct. 28(1971).

U.S. Environmental Protection Agency: Proposed toxicology guidelines. Fed. Register 37(183), 19383(1972).

Books should be cited with COMPLETELY CAPITALIZED name(s) and initials of the author(s), full title, edition or volume, page number(s), place of publication, publisher, and year of publication in parentheses.

Examples:

BEVENUE, A.: Gas chromatography. In G. Zweig (ed.): Analytical methods for pesticides, plant growth regulators, and food additives. Vol. I, p. 189. New York: Academic Press (1963).

DORMAL, S., and G. THOMAS: Répertoire toxicologique des pesticides, p. 48. Gembloux: J. Duculot (1960).

HARTE, C.: Physiologie der Organbildung, Genetik der Samenpflanzen. In:

Fortschritte der Botanik. Vol. 22, p. 315. Berlin-Göttingen-Heidelberg: Springer (1960).
METCALF, R. L.: Organic insecticides, their chemistry and mode of action. 2 ed., p. 51. New York-London: Interscience (1961).

4. Illustrations

Illustrations of any kind may be included only when indispensable for the comprehension of text; they should not be used in place of concise, clear explanations in text. Schematic line drawings must be drawn carefully and clearly. For other illustrations, clearly defined black-and-white glossy photographic prints are required. Should precisely placed indication darts (arrows) or letters be required on a photograph or other type of illustration, they should be marked neatly with a soft pencil on a duplicate copy or on an overlay, with the end of each dart (arrow) indicated by a fine pinprick; darts and lettering will be transferred to the illustrations by the publisher.

Photographs should be not less than five × seven inches in size. Unimportant and indistinct strips or areas on the edges of photographs should be marked on the back of the glossy print (pattern) with pencilled down-strokes, in order that the reproduction surface will not be unnecessarily large; alterations of photographs in galley-proof stage are not permitted. *Each photograph or other illustration should be marked on the back, distinctly but lightly, with soft pencil with first author's name, figure number, manuscript page number, and the side which is the top.*

If illustrations from published books or periodicals are used, the exact source of each should be included in the figure legend; if these "borrowed" illustrations are copyrighted by others, permission of the copyright holder to reproduce the illustration must be secured by the author.

5. Nomenclature

All pesticides and other subject-matter chemicals should be identified according to *Chemical Abstracts,* with the full chemical name in text in parentheses or brackets the first time a common or trade name is used. If many such names are used, a table of the names and their precise chemical designations should be included as the last table in the manuscript, with a numbered footnote reference to this fact on the first text page of the manuscript.

6. Miscellaneous

Abbreviations. Common units of measurement and other commonly abbreviated terms and designations should be abbreviated as listed below; if any others are used often in a manuscript, they should be written out the first time used, followed by the normal and acceptable abbreviation in parentheses [e.g., Acceptable Daily Intake (ADI), Angstrom (Å), picogram (pg), parts per trillion (ppt)]. Except for inch (in.) and number (no., when followed by a numeral), abbreviations are used without periods. Temperatures should be reported as "°C" or "°F" (e.g., mp 41° to 43°C).

Abbreviations

A	acre	kg	kilogram(s)
bp	boiling point	L	liter(s)
cal	calorie	mp	melting point
cm	centimeter(s)	m	meter(s)
cu	cubic (as in "cu m")	μg	microgram(s)
ft	foot (feet)	μl	microliter(s)
gal	gallon(s)	μm	micrometer(s)
g	gram(s)	mg	milligram(s)
ha	hectare	ml	milliliter(s)
hr	hour(s)	mm	millimeter(s)
in.	inch(es)	mM	millimolar
id	inside diameter	min	minute(s)

<u>M</u>	molar	lb	pound(s)
mon	month(s)	psi	pounds per square inch
ng	nanogram(s)	rpm	revolutions per minute
nm	nanometer(s) (millimicron)	sec	second(s)
<u>N</u>	normal	sp gr	specific gravity
no.	number(s)	sq	square (as in "sq m")
od	outside diameter	vs.	<u>versus</u>
oz	ounce(s)	wk	week(s)
ppb	parts per billion	wt	weight
ppm	parts per million	yr	year(s)
/	per		

Numbers. All numbers and fractions or decimals are arabic or roman (table numbers only) numerals. Numerals should be used for a series (e.g., "0.5, 1, 5, 10, and 20 days"), for pH values, and for temperatures. When a sentence begins with a number, write it out.

Symbols. Special symbols (e.g., Greek letters) must be identified in the margin, e.g.,

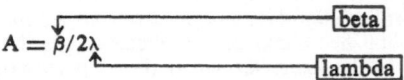

$$A = \beta/2\lambda$$

Percent should be % in text, figures, and tables.

Style and format. The following examples illustrate the style and format to be followed (except for abandonment of periods with abbreviations):

KAEMMERER, K., and S. BUNTENKÖTTER: The problem of residues in meat of edible domestic animals after application or intake of organophosphate esters. Residue Reviews 46, 1 (1973).

The Chemagro Division Research Staff: Guthion (azinphosmethyl): Organophosphorus insecticide. Residue Reviews 51, 123 (1974).

7. Page proof (Galley proof is no longer sent)

Corrected proof must be returned, within two weeks of receipt, to the editor. Author corrections should be *clearly* indicated on proof with soft pencil or with ink and in conformity with the standard "Proofreader's Marks" accompanying each set of proofs. In correcting proof, new or changed words or phrases should be carefully and legibly handprinted (*not* handwritten) in the margins.

8. Reprints

Senior authors receive 30 complimentary reprints of a published article. Additional reprints may be ordered from the publisher at the time the principal author receives the proof.

9. Page charges

There are no page charges, regardless of length of manuscript. However, the cost of alterations (other than corrections of typesetting errors) attributable to authors' changes in the page proof, in excess of ten % of the original composition cost, will be charged to the authors.

If there are questions that are not answered in this leaflet, see any volume of *Residue Reviews* or telephone the Editor (see p. 1 for telephone numbers). Volume 3 (Ebeling) is especially helpful.